INNOVATION IN TRANSLATION

DAVE FERRERA

INNOVATION IN TRANSLATION

HOW BIG IDEAS REALLY HAPPEN

ForbesBooks

Published by ForbesBooks, Charleston, South Carolina.
Member of Advantage Media Group.

ForbesBooks is a registered trademark, and the ForbesBooks colophon is a trademark of Forbes Media, LLC.

Printed in the United States of America.

10 9 8 7 6 5 4 3 2 1

ISBN: 978-1-946633-84-2
LCCN: 2020919419

Cover design by Carly Blake.
Layout design by Megan Elger.

This custom publication is intended to provide accurate information and the opinions of the author in regard to the subject matter covered. It is sold with the understanding that the publisher, Advantage|ForbesBooks, is not engaged in rendering legal, financial, or professional services of any kind. If legal advice or other expert assistance is required, the reader is advised to seek the services of a competent professional.

Advantage Media Group is proud to be a part of the Tree Neutral® program. Tree Neutral offsets the number of trees consumed in the production and printing of this book by taking proactive steps such as planting trees in direct proportion to the number of trees used to print books. To learn more about Tree Neutral, please visit **www.treeneutral.com**.

Since 1917, Forbes has remained steadfast in its mission to serve as the defining voice of entrepreneurial capitalism. ForbesBooks, launched in 2016 through a partnership with Advantage Media Group, furthers that aim by helping business and thought leaders bring their stories, passion, and knowledge to the forefront in custom books. Opinions expressed by ForbesBooks authors are their own. To be considered for publication, please visit **www.forbesbooks.com**.

To the next generations of innovators, whose brilliance and resilience will shape the future of our blue marble. And to my kids, Conor, Natalia, and Emilio, who teach me daily about how to break through, stay focused, think forward, and keep on traveling.

CONTENTS

ACKNOWLEDGMENTS 1

FOREWORD BY OREN KLAFF 3

INTRODUCTION 7
Innovation Is a Team Sport

CHAPTER 1 27
Find the Right Solution for the Right Problem

CHAPTER 2 49
Own Your Solution

CHAPTER 3 77
Build It, and They Will Come

CHAPTER 4 107
Every Success Is Built on Failure

CHAPTER 5 123

How the Money Moves: Fundraising Essentials for Fledgling Entrepreneurs

CHAPTER 6 147

Assemble a Winning Team

CHAPTER 7 173

Planning and Executing a Limited Market Release

CHAPTER 8 187

The Xtreme Stress Management Guide for Entrepreneurs and Innovators

EPILOGUE 217

My Blue Marble

In the end, all business operations can be reduced to three words: people, product, and profits. Unless you've got a good team, you can't do much with the other two.

—Lee Iacocca

You miss one hundred percent of the shots you never take.

—Wayne Gretzky

ACKNOWLEDGMENTS

I am happy to say that *Innovation in Translation* lives up to its title. This was very much a team effort. I was inspired by Oren Klaff to write this "how-to." I was reading *Pitch Anything*, which led to our friendship and also set in motion a lot of the thinking that led to this book. Thank you for everything, Oren. I'm so glad you are a part of my life.

Thanks also to Bev West. This book would not have been written were it not for her literary expertise. We spent many Sundays on the phone, telling stories about innovation and about life. We've also become friends and kindred spirits. She's a word wizard and a master storyteller.

I owe thanks most of all to my parents, Dick (rest in peace) and Louise. This is a book about the meaning of work, and I learned that work can be meaningful from my father. Everything he did—from his most complex mechanical designs to digging in the garden and coaching my team sports—he tackled with joy and resolve and enthusiasm. My earliest memories of my father are of seeing him work and coaching, all the while realizing that he was happy. I did not know it then, but that was one of the most precious gifts a father can give his child. My mother, for her part, taught me how to be patient and trust myself, two priceless gifts that have aided and supported me throughout my life.

Many, many thanks to my friends. As always, I prevailed upon them to critique various drafts of the manuscript. Happily, they complied, and *Innovation in Translation* is infinitely better as a result.

Thanks also to Keith Kopcsak, Y-Danair Niehrah, Nate Best, Jack Reich, Carly Blake, and all the wonderful folks at ForbesBooks for their careful and thoughtful stewardship of this project.

Thanks to my sister Diane and brother-in-law Paul, for encouragement throughout the years and and keeping family life fun.

Finally, a very special thanks to Kim, my life partner, who has supported and inspired me throughout this process to continue on and to complete this educational book.

FOREWORD BY OREN KLAFF

'm an investment banker. It's a rough business. Everyone wants money to grow their business, but only a few will ever get it. I love to help people with brilliant ideas to connect with the resources they need to grow a strong, healthy company. And if you do everything that Dave tells you to do in this book, eventually you will be speaking to me, or a person just like me, when it comes time to get serious and raise some money.

I talk to people with great ideas every day. People with new and life-changing innovations. Entrepreneurs who are passionate visionaries. Engineers who are as bold as they are smart. So how many deals do I do a year? Not many. Most of these entrepreneurs, engineers, and founders are 100 percent not bankable. At the end of a meeting, my team and I are scratching our heads because we know that even though the product is amazing and groundbreaking and potentially very rewarding, nobody but nobody is going to give this person any money. Seriously. Not even five bucks.

Investors aren't paid to be innovative; they're paid to create profits. Think of it this way: nobody got control of $200 million or a billion to invest without being smart, aggressive, and ruthless. Today, investors mean business and they have no weaknesses or "blind spots." As an inventor and an entrepreneur, it's easy to convince yourself that your new product is so good, so revolutionary that everything else that you need will just fall magically into place. When you've just solved an unsolvable problem, why should you have to worry about investors and sales and marketing or manufacturing relationships? These things will just come to you because the value of your product is so obvious.

The only problem is, no they won't come. Don't confuse the difference between a product and a company. Most people do, but you shouldn't. Products don't have "management teams" or attract reputable board members. Companies do. Products don't have partnerships, pursue regulatory approvals, or get financed. Companies do.

Listen, I get it. Creating and perfecting a revolutionary new product is fun and exciting. While at the same time, supply chain management, staffing, HR, fundraising, and accounting feels like a death march. Understandable. This is a problem that I have myself. I love new ideas and engineering breakthroughs, but most of the time, I have to work with some pretty boring, straightforward, basic, step-by-step kind of people to get the job done. They aren't running at my pace; they don't share my vision and they are never going to.

And they don't have to. *They just have to believe in me and do a great job at doing their job.*

Every morning, I try to remember that there is a crucial place in any new venture for those people who "don't get it" and share my vision for a new and better world. You sometimes feel like it's sabotage. All they want to do is slow down the process, put systems, structure, and critical business elements in place. But if you want a company that can grow and dominate, yes, you need to crush barriers, break down doors, and lead from the front. But … you also need credibility, which leads to money; you need people, which lead to growth; and you especially need those lawyers for those times when you move a little too fast and step over the line. You can't change the world by yourself. Without a team around you, *nobody is going to fund you.*

Speaking as the guy who is in the position of selling your big idea to big money, you can only get away with introducing one really novel thing to the world at a time. That new thing is your product. If people are going to feel comfortable enough to jump in the boat with you, you have to surround your breakthrough with the trusted and the familiar.

In other words, every brave new idea still has to fit into a traditional financial and organizational infrastructure. You have to look a lot like the other companies that are succeeding in your industry. Accounting. Financial controls. Legal. Advisers. Intellectual property. All that needs to look boilerplate and standard. Where engineers and entrepreneurs always get sideways is that they start believing that they're playing a new game when the old rules still apply, maybe even more so.

Whether it's true or not, here's what every investor believes in his core: *Ideas are cheap, everybody has them*. Even big ideas are not worth anything until somebody invests in it and customers line up to buy it. Dave's book is a road map through this world. It's a world you will have to navigate in order to bring your big idea to life. Dave's book is a step-by-step guide to commercial success in innovation that every engineer and entrepreneur can understand. I'm putting a copy of it in the hands of every new engineer I meet, knowing I've improved their chances of success a hundred-fold.

—**Oren Klaff**, Investment Banker, Managing Director, Intersection Capital and Best-Selling Author of *Pitch Anything* and *Flip the Script*

INNOVATION IS A TEAM SPORT

Picture this. It's 1998. I'm in Buenos Aires. I had formed a team a year earlier to develop a cerebral aneurysm coil embolization device that would give doctors the opportunity to treat hemorrhagic stroke in an improved way. It had been a year of high-pressure engineering to prepare a prototype. We had designed, built, and tested thousands of these devices in bench models, all leading up to this. Would the device work in people? We were about to find out.

First-in-man studies are set by the doctors, so we had no control over our timing. Our first study had been set by Dr. Pedro Lylyk, a world-renowned doctor in Buenos Aires, and we had no choice but to move and move fast. So, we scrambled to build three hundred devices of various sizes before the deadline. The devices had just come back from the sterilizer two days before. Under ordinary circumstances, these would have been made and ready weeks in advance for me to ship according to regulatory and customs laws. But these weren't ordinary circumstances. I had to hand carry all three hundred devices myself into Argentina.

I packed three duffel bags with one hundred devices in each one. I checked these, along with all of my luggage, and took off for Ezeiza International Airport. I looked like a local, or at least that was the plan. I'm Italian as a lot of Argentinians are. I was hoping I would just blend in. I knew that hand carrying three hundred sterile coils into a foreign country was a dicey plan. I was playing it low key, trying not to attract any unnecessary attention until I made it through customs.

I picked up my bags from the carousel, put them carefully on a push cart, and got into a line. It didn't really matter which line I picked, because customs is a crap shoot at the Ezeiza airport. There isn't a "declare or nothing-to-declare" system. Everybody pushes a button, and randomly the machine will illuminate either a red or a green light. If the light is green, then it's your lucky day and you can exit the airport without getting your bags searched. If you get a red light, then you have to load your bags onto a conveyor belt to pass through an X-ray machine and go through customs. I was lucky that day. Green light. Awesome.

I was smiling, basking in the sunlight of my good fortune and heading for the exit. Then, out of nowhere, a customs agent appeared, looking me right in the eye as he approached. I kept smiling. I mean,

I got a green light—what did I have to worry about? Soon I would be putting up my arm to hail a taxi to take me to my hotel in Plaza San Martin. Later, I'd have dinner at my favorite Italian restaurant, Dora, a few blocks away. I could already smell the Bolognese sauce. Life was good.

Just as the doors swung open to the outside, the customs agent said something to me in Spanish. I don't speak Spanish, and that's exactly what I told the guy.

"*Lo siento pero*. I don't speak Spanish," I said.

"How was your vacation?" He asked me in heavily accented English.

"I'm not back from vacation," I said. I mean, I wasn't about to start lying to customs agents. I wasn't that crazy. But I was flattered he thought I was local. My low-key approach had worked. So why was I here? "I'm here on business," I muttered.

"Why do you have all these bags?" he asked, and I shrugged, because if I told him the truth, I knew there was going to be trouble. Trouble was exactly what he was looking for.

"Come with me, please," the agent said. "And please bring your bags." The agent led me and my overloaded cart back to the inspection area and brought me to the front of the line, pushing past all of the unlucky red lighters still waiting in line. Then he told me to load all five of my bags onto the X-ray conveyor. After the bags passed through the X-ray, he took one of the duffels and zipped it open—to reveal one hundred neatly packed devices. He looked up at me and winked. Then he unzipped the other two duffels. Uh-oh.

"What are these?" he asked, and now he was smiling—and it wasn't a friendly smile. Although he did seem happy about something, and I sensed that something would be at my expense.

"Medical devices," I said, but I already knew I was screwed. He was just playing with me now. And he was enjoying himself.

"How many of these devices do you have here?" the agent asked, looking over the three bags.

"Three hundred," I said, because now there was really no purpose at all to lying. I was already caught.

"Why didn't you declare these?"

"Because they have no value," I told him.

"Everything has value," he said and looked at me meaning- fully. I began to see where this was headed. He was smiling. I was sweating even though it was June, which is the middle of winter in the southern hemisphere. We just stared at each other, waiting to see who was going to blink first.

"How much are these worth?" he asked me. I blinked. Well, it was Argentinian customs. He had all the power, and no matter what I said at this point, I was going to wind up giving him the answer he wanted eventually. May as well cut to the chase.

"I told you, they're worth nothing," I said again, only louder, as if more volume would make him understand me better. "These can't be sold, so they have no value except to me." The agent zipped up the bags, asked me to load them back onto the push cart, and told me to follow him. He opened the door to a small conference/interrogation room and told me to sit down. Then he started to shake me down.

Why was I in Argentina? Who was I meeting here? How long was I staying? What did I intend to do with these devices? Where was I staying? And on and on and on. About an hour in, he got down to brass tacks. He told me I had three choices:

1. Fly back to the US tonight with the product.

2. Allow him to open all the packages and break the sterile barrier, making them useless for any clinical trial.

3. Pay the tax. And by *tax*, I was pretty sure he meant a bribe.

"How much is the tax?" I asked.

"Ten dollars per device, so three thousand dollars," he said, and it didn't sound like he was open to haggling.

"I don't have that kind of cash on me," I said, but it didn't seem to make a difference to him. He got all Judge Judy on me, repeating my three choices over again, like I was a shady character ducking out on personal accountability … and also pretty slow on the uptake.

"Can I pay you in a couple of days? I'm not from Argentina, so I will need a couple of days to access cash like that." This, I realized instantly, was a pretty dumb suggestion. Once I left this airport, his leverage would be gone. He knew that. But he fell for it anyway. Maybe he knew something I didn't. This worried me.

"You have twenty-four hours," the agent said. "And until you return, we will hold on to these." He put his hand on my bags and opened the door for me to leave. Ah, there was the catch. Well, there went that great idea. While I was relieved to get out of that airless interrogation room, I really had no idea how I was going to put my hands on three grand in twenty-four hours. I really, really didn't want to leave my bags behind, but I had no choice.

"Okay, see you tomorrow," I said and practically ran to the cab stand. While I was waiting, I called my distributor in Buenos Aires and asked him if he could loan me $3,000 to get the products out of customs. He laughed and said he couldn't. He said he knew this was a bribe and didn't trust that my CEO would pay him back. I didn't blame him.

I got into a taxi and headed for the Marriott Hotel in the Plaza San Martin. While checking in, I asked the front desk person if I could get a cash advance. She smiled and said she could do this for me, but when I mentioned $3,000, she shook her head and told me her limit was a hundred dollars a day. Strike two.

I took the room key and was headed toward the elevator when the bellman approached me. The bellman turned out to be Juan. I knew this kid well, because I'd been back and forth to Buenos Aires fifteen or twenty times over the last eighteen months, and I tip well.

"Mr. Ferrera," he said politely. "I couldn't help but overhear that you requested a three-thousand-dollar cash advance." His English is perfect because he grew up in Texas. He moved to Argentina when he was thirteen so his dad could work at the Ford Motor plant in Argentina. That was only five years ago. This gig at the Marriott was his first job, and he was a good kid, so they loved him here. I was very happy to see Juan. He always had answers I couldn't have thought of myself.

"Yes, Juan, I need funds to pay a tariff at the airport that I didn't expect. It hasn't been a very lucky day for me."

"Do you have an American Express card, Mr. Ferrera?" Juan asked me, and I told him that I did. "Then your luck has just changed," Juan said, and he flashed me his signature gap-toothed grin that earned him a lot of tips, I was guessing. Juan told me about a new program he'd heard of called Ameri-Cash that would link your bank account to your Amex card and allow you to withdraw unlimited funds as long as the money was in your checking account.

"Thank you, Juan," I said. I tipped the kid almost all the cash I had left in my pocket, went to my room, and called Amex to check out if there really was such a program. And miracle of miracles, there was. I set up an account over the phone and made a mental note to hire Juan the first chance I got. I like having people around that have solutions I can't come up with myself. I went back downstairs, slipped Juan some more green, and asked him to line me up a security guard. An hour later, I withdrew four grand—an extra grand just in case—and later that afternoon I headed back to the Ezeiza airport to pay off my new best friend, the customs agent.

Once inside, I spotted my buddy right away. He did a double take. I smiled, put my hand in the air, and rubbed my fingers together—the international symbol for payola—and it spooked him. He hurried over to me and rushed me back behind closed doors. Once back in the conference room, I pulled out thirty one-hundred-dollar bills and handed them over. He took the money and counted it. Then, he counted it again. Then he told me to wait and left the room.

Two hours later, he still hadn't returned, and I was really starting to sweat. So many thoughts were going through my head. I had just paid a bribe. Was I going to get arrested? Had *he* been arrested? Was he going to ask for even more money now? Where were my bags of devices? Why was I so hot in the middle of winter?

Finally, he returned, carrying my bags. He dropped them on the floor. I grabbed them and asked for a receipt for my payment so I could be reimbursed. He looked at me like I was born yesterday.

"I never want to see you in my airport again," he said. That was going to be tough, since I was flying out of here in three days. I decided not to mention it. Best not to push my luck. I did not get a receipt.

The next day, I stood by the table as Dr. Lylyk performed the first-in-man neurointerventional procedure with the devices I had just bribed out of customs, and the operation was a complete success. Once again, my ability to understand what people are trying to tell me from the other side of the dilemma had led me to a solution that worked for everybody; the doctor, his patients, Micrus, and even the surly and definitely corrupt customs agent were pleased. This is really what innovation is about. It's about great ideas and new ways of thinking through a problem for sure. But it's just as much about getting people to work together as a team.

Why Today's CEOs Need to Know How to Coach Sports

Although you'd never guess it from the story I just told you, I am not a secret agent or an international arms trader. I am an engineer. People have always been a little confused by me, because even though I'm an engineer, I don't act like one, and I never really did. I defy all the stereotypes.

I was a jock in high school. I played baseball and hockey, I never watched *Star Trek*, and I did not spend my Friday nights hanging out with my friends and hooking up stuff to see if it worked. Yet I can geek out for hours about the fascinating history of interventional neuroradiology, and I read the newsletter from the US Patent and Trademark Office just for kicks.

I'm a hybrid. And while that was kind of difficult in the beginning, it's turned out to be a very useful quality, because being an innovator is about a lot more than inventing stuff. I'm an engineer for sure, and I anchor myself in that ability to see solutions within problems through that lens. But I'm also the guy who has to convince other people that my idea is a good one and ask them to invest. I'm the guy who has to hire the team that will arrive at a marketable product within budget.

I'm also the guy who has to bridge the death-defying canyon that divides entrepreneurs and engineers. I'm the guy who has to explain to the doctors how the device works and why it's a good idea for them to be early adopters. And I'm the guy who has to instruct them, sometimes mid-procedure, of what to do when the device unexpectedly fails. I'm also the guy who has to tell the entrepreneurs to settle down and wait for a product that works consistently, no matter how badly they want to go to market. And I'm the man on the wall who has to keep these two differently motivated teams, the sales team and the engineering team—the offense and the defense—to resist the urge to kill each other and instead focus their energies on our collective goals.

The battle between entrepreneurs and engineers is ancient and primal. Our adversarial relationship is practically embedded in our DNA. When I'm trying to build consensus around a conference table of engineers and entrepreneurs, I'm fighting centuries of generational hostility that probably dates back to the guy who first invented fire and told his neighbor, who went and filed the patent behind his back.

I guess you could say that in addition to being an engineer, I'm a coach who sees everybody's strengths and everybody's weaknesses and puts them together into a team that is symbiotic and committed and equal to the problem. I have to say, after all of my years in this business and playing on both sides of the game, I've become a pretty good coach. I think this is because I had a couple of great coaches myself growing up, starting with my dad.

My father was a tool and die man. He was so gifted at drafting and engineering design. Growing up, I wanted to be just like him. I admire people that have a special talent that brings them the

> The battle between entrepreneurs and engineers is ancient and primal. Our adversarial relationship is practically embedded in our DNA.

kind of joy and excitement my dad felt when he was working. My dad was passionate about plastics and one of the best plastic mold makers in Leominster, Massachusetts, and probably in the world. He was proud of that. So was I.

After plastics, my dad loved coaching baseball and hockey. So, of course, I played baseball and hockey. He coached most of my teams from the time I was old enough to stand up on skates. His dream for me, though, was to play professional baseball. Thanks to an injury while I was in college, things didn't work out that way, but if he was disappointed about that, he never let me know. I think he felt I got what he wanted me to get out of sports, so it didn't matter that I became an entrepreneur. He knew that even though it was a different game, he could be sure I was playing by the rules of fair play that he'd taught me, and that was what was important to him.

As a coach, my dad stressed the value of teamwork. He taught me and every kid he coached to be a good team player, to be coachable, to care about our teammates, and to put team before self. As an engineer, he taught me the joy of making something, of solving problems that couldn't be solved, of creating things that hadn't existed before. He taught me to see solutions in the world rather than problems. In short, he gave me the seemingly incongruous tools that I needed to succeed in a future he hadn't yet imagined.

Debunking the Myths about Innovation

I've been part of several inventions, but the one BIG invention I was part of was a medical device called the stent retriever, which doctors can use to treat ischemic strokes. Before this device, doctors had no effective or reliable way to intervene to remove blood clots in the brain, and so their patients were either severely debilitated or they died, and there was literally nothing the doctors could do about it.

I co-created a collapsible cage that allowed doctors to go in and grab the blood clot and remove it from an artery, allowing the blood to flow freely once again to the brain. With this one simple development, people could survive an ischemic stroke. Their doctors were no longer powerless to intervene. It was an incredible feeling to be at the helm of something that really made a difference in people's lives. But it didn't come easily. And I didn't do it alone. Innovation is a team sport, and champions are forged over time.

History has told us a lot of fairy tales about how new ideas enter the world. An apple falls on Sir Isaac Newton's head, and *boom*! Gravity! The Wright brothers watch a bird in flight, and suddenly we have the biplane. This isn't the way innovation works at all, even for the Wright brothers, who had labored and failed for years before achieving flight and were only one of a few teams around the world who were developing the technology to fly. There is nothing sudden or solo about innovation. It takes years of thoughtful development, testing, and teamwork to really make a new idea happen in the world. Without a team, there is no progress, and this is true of everything from the airplane to the telephone to the personal computer. Progress takes a village.

There is a myth that inventors are the geniuses who generally get ripped off by history because they are too naive to protect themselves and their inventions from profit-hungry businesspeople who only want to exploit the inventors' genius for their own greedy and nefarious ends. The Grimms' fairy tale version of innovation persists today, even at the highest levels of business, where billion-dollar

> There is nothing sudden or solo about innovation. It takes years of thoughtful development, testing, and teamwork to really make a new idea happen in the world.

inventions are on the table. And nothing about this myth is true. It's completely phony. It takes engineers AND entrepreneurs to make big things happen. You can't have one without the other.

As Americans, we are especially biased when it comes to innovators, because we are a nation of innovators. Innovation is a uniquely American trait. We're a country built on brand-new and controversial ideas about everything, including our government. And we've been taught since grade school that the people who came up with those ideas, our founding fathers like Thomas Jefferson, Benjamin Franklin, or Alexander Hamilton—all part-time inventors and engineers, by the way—are the heroes of our story. But this is only half the story.

America's story, just like the story of every other big innovation that changed the world, is also in large part a business story. Along with a struggle to bring a new idea into the world, the American Revolution was a business play on the part of the founders of America to take back control of their invention from the parent company who owned the patent. The first colonies were corporations, and Britain stepped in and took ownership only when those businesses went bankrupt. As a result, we had to turn to new investors and lenders to preserve our private ownership.

The inventors of America didn't break off from the Brits and build America all by themselves. They partnered with the French and the Spanish and the Dutch. Without these strategic foreign alliances, without the assistance of business leaders and venture capitalists around the world, we would have lost the revolution.

This is the other part of innovation that I love so much—it's international. Big ideas don't recognize borders. We are all part of the global team of new knowledge, new experiments, all working together to solve the same common problems, each with their specific skill sets, all laboring to make the world a better place. When you undertake to become an innovator, you join that global team and take up your small share in the advancement of human knowledge.

Let's get started. The world needs you. Right now.

Are You a Santa or an Elf?

Before you begin any new project, it's important to understand where you fit into the innovation process. While you may be the person who had the brilliant idea in the first place, that doesn't mean you know how to lead a company, or build a team, or build a prototype.

You can't be all things to all people. You need to know what you're good at, and then you need to know how to hire the best people to fill the positions that you can't cover. The very first questions you should ask yourself—before any of the other trillion questions you will need to ask and answer before you get to the end of this process—is, "Am I the person who should lead my new company, or do I need to hire a leader who can take my idea and run with it?"

Here are the questions that I ask all my inventors to ask themselves to determine what position they will play on my team—and whether or not they are CEO material or should focus more on the lab and operations.

1. **Are you Santa or an elf? Are you a leader, and do you take tasks and get them done?** Entrepreneurs manage a wide variety of tasks as part of the business, from engineering and manufacturing, to sales and marketing and accounting, to customer service and more. Can you wear multiple hats, as Santa does with Christmas, or do you prefer to be the elf who loves to execute specific tasks? Do you take the initiative, or do you want clear instructions? Santas make better entrepreneurs than elves do.

2. **How do you handle and manage money?** Starting a business requires money at the beginning, both to operate it and for you to live on while it scales. Will you be relying on your own money, investors' money, or both? If you are a big spender and aren't great at managing money, those bad habits are likely to follow you into a business. And if you are usually unable to make worthwhile investments in the future of your business for fear of ending up living in a cardboard box, then you should probably put somebody else in charge of the money. People who develop solid, unconflicted, and unemotional relationships with money make better Santas.

3. **Do you need to know exactly where you are going, or are you okay flying blind and on instinct?** The only thing that is certain in business is that nothing is certain. The only

constant is change. Are you comfortable with being uncomfortable? Can you handle taking educated risks and surviving the constant ups and downs of owning a business? If you are looking for certainty or a drama-free zone, you may find yourself terrified of the entrepreneurial roller coaster. If so, don't become a Santa—you will be miserable, and your business will suffer as a result.

4. **Are you able to make and keep your commitments?** Running a successful business is not just about having great ideas. It's also about strong execution sustained over time. So, if you have difficulty staying focused, if you are lousy with commitments and averse to the idea of working day in and day out on the same thing, then become an elf, not a Santa.

5. **Were you born to create a business?** Were you interested in business as a child, or did your family run one? Did you seek out entrepreneurial roles in school, in social organizations, or even in your previous job? A natural inclination or past interest in entrepreneurship seems to be a good potential indicator of future success.

Every innovation requires both Santas and elves. Neither is more essential than the other. The important thing is to be realistic and fact based in your assessment of your skills, then assume a role you are comfortable with. Otherwise, this process will not be a positive experience for you in the long run.

CHAPTER 1

FIND THE RIGHT SOLUTION
FOR THE RIGHT PROBLEM

*Never mess with Mother Nature, mother-in-
laws, or motherfucking Ukrainians.*

—*The Italian Job*

W hen I was thirteen, I was a shortstop on my Babe Ruth
Baseball League team. Our pitcher was a kid named
Scott Young, who had the world's fastest fastball. At
thirteen, he could throw a ball eighty-five miles an
hour. It was unreal. Nobody could get a hit off that kid. We should
have been unbeatable that season. Scott outmatched everybody on
any team we played. The problem was, he also outmatched his own

27

team. Nobody could catch his pitches, and so the catcher would drop the ball or signal wild pitches out of fear, and we'd lose games.

We had a strange problem. Our pitcher was TOO good. What could we do? It wasn't like the coach could tell a kid like Scott, "Hey, slow your roll. Stop being such a prodigy." Scott's incredible talent was a problem that had to be solved creatively. The coach needed to innovate his way around his pitcher's freakishly magnificent arm. And that's the day I went from being a shortstop to being a catcher. The coach had seen me interact with Scott's arm. He knew that when Scott gunned a ball at me, I held on to it. So, he moved me into the catcher position, Scott pitched like the star he was, I wasn't afraid and could catch the balls he threw, and we ended up winning the state championship that year.

I learned some important lessons about problem solving from my coach that season. Scott's talent was a problem that couldn't and shouldn't have been solved. But still, we had to win games. Coach came up with a solution that turned a liability into an advantage without ever addressing the source of the problem directly. He solved a pitching problem with a shortstop. Coach taught me that problems can be solved from a variety of angles. By the way, Scott ended up making the pros, but in ice hockey. He had a seventeen-year career in the NHL.

This understanding is at the heart of my rule number one in innovation. Not every problem requires an immediate solution, and some problems, like Scott's arm, shouldn't be solved at all. It was the problem with the people around him that needed to be addressed. Successful innovation begins with picking the RIGHT problem at the right time—and then matching it with the RIGHT solution. When all of these elements align, you get groundbreaking innovation. When these elements do not align, you get things like the eight-track tape.

The best way to enter the field of innovation is to enter through the lens of your personal or professional passions. Find a problem that really means something to you—something that you know about, care about, and understand thoroughly. Or, as was the case in my situation, be alert to the unsolvable problems that find you, for which you already have the knowledge to confront. If it turns out to be a problem that a lot of people grapple with, if the pain point radiates far beyond you, then you may have run into a problem that is begging for a commercial solution.

> Successful innovation begins with picking the RIGHT problem at the right time—and then matching it with the RIGHT solution.

My *Eureka!* Moment

When I was twenty-one, my dad called to tell me that my grandfather, my Pepe who had been at every single one of my baseball and hockey games since I was five, was in the hospital. I was living in Lowell, getting my engineering degree, playing college ball, and interning for a medical device company called USCI in Billerica, Massachusetts. I was due to leave the next day because my sister was getting married over the weekend. My dad told me that Pepe had undergone a routine carotid endarterectomy to remove plaque from his carotid arteries. It's a bread-and-butter procedure for any neurosurgeon worth his salt. But something had gone wrong during the procedure, and my pepe had suffered an ischemic stroke.

An ischemic stroke occurs when a major artery supplying blood to the brain gets blocked. That blockage is called an occlusion, and if the occlusion is not removed, the brain cells become deprived of oxygen, and eventually the brain dies—and so does the patient. My dad said Pepe wasn't going to recover but that if I came home right away, I'd be able to see him. I hopped in the car, gunned it to Leominster, and picked up my dad, and we headed straight for the hospital. Pepe was sedated, talking to God, asking for divine intervention. He was Italian, so he put his right fist in the air as he talked, just in case God wasn't paying attention.

I knew that he had had a stroke on the right side of his brain. I knew this because he was still able to yell at God. Strokes on the left side of the brain impact speech. Strokes on the right side of the brain cause confusion, so I knew Pepe really didn't understand me or what was happening to him. That was probably a good thing. I hoped he realized that I was there, but I couldn't be sure.

It was very hard to see the old man like that. I loved him so much and was just getting to the point in my life where I could really appreciate what a huge role he'd played in my life. In those days, I had been making it a point to drive home from college and share a pot of pasta and meatballs with him and talk but mostly listen to my grandfather. He would tell me how fascinating it was to have been born when he was—to see the emergence of so many world-changing inventions: the automobile, radio, TV, air travel, space flight, and even personal computers, which totally freaked him out and yet delighted him too. He asked me what I thought I'd see developed in my lifetime. I never had much to say at these moments. What the heck would I ever see or do? I played ball.

The neurologist came over and began talking to my dad. I heard him say that the ischemic stroke would most likely transform into a hemorrhagic stroke, which would kill Pepe. The doctor told my dad that all we could do was wait.

"And what are you doing with yourself, young man? Putting your time to good use, I hope?" the neurologist said, unexpectedly putting all the focus on me. I wasn't happy about becoming the center of attention just then. I was feeling a little tongue tied, and that's always scary for me, since I stuttered as a kid. I think I was a little in shock at seeing my grandfather so stricken, and it seemed like a weird time to be quizzing me on my accomplishments. I just stared at him.

"Are you going to college?" the doctor prompted me.

"Sorry, I think I'm still in shock," I said, pulling myself together. "I've never seen anybody in this condition before. Yes, I'm in college."

"What are you studying?" the doctor asked me, and all of a sudden I had the feeling he'd gotten some advance press on me from my dad.

"Engineering," I said. "I'm specializing in plastics." The neurologist nodded, as if I'd given him the answer he was looking for. "What can we do to help my grandfather?" I asked the doc, because just waiting for Pepe to die didn't sound like a very good treatment option.

"Well, as an aspiring engineer in plastics, there is one thing you can do," the doc said. "You can study hard, get great at what you do, and then you can invent a device that would allow me to treat ischemic stroke patients like your grandfather, because right now, I don't have the tools I need to intervene. There's nothing I can do for him right now. But I sure would like to be able to do something to help in the future for families just like yours."

While that didn't really sink in at the time—I was probably too pissed that he was telling me there was nothing he could do for my grandfather—several years later, the doc's words came back to me. I followed his advice and wound up coinventing a medical device, the stent retriever, to treat ischemic stroke patients with large-vessel occlusions, just like my Pepe had.

The device looks like a small metal cage attached to the end of a long wire. The cage is collapsible, which makes it possible for a doctor to go in and grab blood clots or any occlusion in your brain and remove them from an artery or vein to allow blood to flow freely once again. If a solution like this had been available eighteen years earlier, my Pepe would probably still be alive.

While it's unfortunate that I was too late for my Pepe, I know that this device has helped lots of other people's loved ones since then, and that makes me feel really great about what I do for a living. This was my introduction to the power of innovation. I realized that I had the ability to solve problems that had caused pain to a lot of people. An invention that I had spearheaded was saving lives. I and my team had solved a real problem that touched a lot of people.

The Problem Quiz

Just like in the old adage "One man's trash is another man's treasure," things that you perceive as a problem might not seem like much of a problem to anybody else. Picking the right problem, at the right time, for the right group of people is a critical preliminary step in successfully bringing an innovation from idea to market. If you solve a problem that very few people perceive as a problem, you won't have a market. But it can be hard to get perspective on that when you're on fire with inspiration at the beginning of the game.

For this reason, I have developed list of questions that I call the problem quiz. I ask myself and my team these questions every time I consider whether or not to invest time and money into solving a problem.

1. Is this a real problem or just a perceived problem? Remember the *Curb Your Enthusiasm* episode where Larry meets a guy who invents a car periscope to see the traffic ahead of him to know what's causing it? Not really a big problem. Stupid problems create stupid solutions. Don't make the same mistake! Be honest with yourself, and ask others to keep you honest. How do they feel about the problem? Have they wished there was a solution? Really?

2. What is the pain that the problem is causing people, and how many people are experiencing pain as a result of this problem? Is it just you and your neighbor, or is it the whole block? The broader the pain point, the better the chances that you have identified a problem that should be solved.

3. Is it a problem that can be solved quickly, or will it take years of development?

4. Has anybody else tried to address this problem in the past, and if so, how, and what was the result?

5. If you are a man, have you asked a woman what she thinks of your problem? If you're a woman, have you asked a man?

The Solutions Quiz

In the companies I've cofounded, and now in my venture studio, people come to me with solutions for viable problems all the time, and I have to evaluate them. That's my job, and I have to make these kinds of decisions every day. As a result, I have a checklist that I apply to every solution I encounter. Here are the questions I ask myself and the team:

1. Is your solution something that people who are experiencing the problem could devise, or will it require an expert?

2. Does your solution work with currently available technology, or will you have to wait for future innovation to execute your solution? You want to make sure that you have the means at your disposal to execute a solution now. If it's a long-term solution that requires something else to happen or some new technology to be invented before you can proceed, then you will want to take a different approach.

3. Is your solution cost effective? Will it demand resources far beyond your means or take years to earn out? You and your investors want to know at the beginning an approximate length of process. How much time do you anticipate that it will take for you to get to market? How long will it take to repay your investors? And remember: patents only last for twenty years.

4. Has anybody else come up with your solution, and if so, what was the result? It's important to research the pioneers in your field who have come before you and learn from their successes and their failures. It's also important to research the state of the art in your field currently and identify any potential competitors who may already be in development on the same invention. If there is already a similar product in development, chances are you won't beat it to patent—which is not a definite deal breaker, but it's close.

5. Is your solution reducible to practice that is simple and straightforward? For most problems, the easier the solution, the safer the solution for the end user. I stay away from overly complex solutions that are going to require a huge educational ramp-up before they can be put into practice. People shouldn't need to get an advanced degree in order to put your solution to good use.

Choosing the Right Flight Path

There's more than one way to fly to Atlanta. At the beginning of an innovation development process, there will often be multiple routes to explore before finding the right flight path to get you to your destination. When the team and I at MindFrame began to imagine a solution for ischemic stroke, a problem that affects 2.5 million people around the world, we, along with Dr. Cragg, found that we had two viable approaches: one passive and the other active.

Here is a breakdown of our two possible flight paths:

- **Passive:** The passive approach for us involved a device that would be introduced into a catheter and placed across the clot passively, which could then be activated by the physician mechanically on demand. In addition, the force of the activation would be controlled by the physician with a wire that could be pushed and pulled to open and close the device against the clot.

- **Active:** The active approach involved a device that was made from a material called nitinol, a nickel titanium alloy, which would be self-expanding and always activated. We were very familiar with this material because it is frequently used in medical devices that we had worked with before, so we felt comfortable using it in an active approach that did not require physician manipulation to engage.

We chose the second—the active approach—for ease of design, manufacturing, and assembly. Our experience with this technology gave us confidence that it was more viable, had fewer moving parts, and was simpler overall across the board. It would also be easier to use, was more or less idiotproof in operation, and did not require much training for doctors on how to use the device, which was a time and money saver for us.

We developed this technology within months of forming the company in 2007 and put it into clinical use within a year. The patents were issued four years later, and the company was acquired by Covidien in 2012. Years later, the passive approach was developed by an Israeli competitor and gained market approval. The active approach is still the market leader today due to its ease of use and success rate.

You Don't Need the Whole Highway—Stay in Your Lane

When you are thinking like an innovator and an entrepreneur, problems and solutions present themselves all the time. When you're a serial entrepreneur like I am, you are literally presented with dozens of innovations every year. How do I determine which inventions to get involved with? First rule of thumb: I stay in my lane.

As you can probably already tell, I'm a firm believer in the idea that you can't be all things to all people. You can't solve every problem under the sun. I only innovate and try to iterate in arenas that I understand completely. I stay in my sweet spot. There are plenty of problems to solve right in your area of expertise, and you will only get better the longer you stay and master one arena. Also, there are lots of innovations to be made in relation to your original invention. The longer you stay with a problem, the better your solutions get, and you can generate new innovation by refining and improving your original technology. This too is innovation.

Decide what you're passionate about, learn the area, and then stay there as long as you can. This allows you to build upon a solid foundation that only strengthens over time. This is true when I'm innovating my own solutions, but it's also true when I'm deciding what I will or will not invest in.

I have gone on to invent or improve upon a multitude of solutions all in areas of interventional neuroradiology, but I know nothing about water redistribution. This is why I passed on a deal that came across my desk, even though it looked promising. I do this with lots of deals every day that look good but lie outside my expert zone. I stick to the problems that I understand. If it's outside my area, I don't get involved. Other people who know the science will do a better job than I could. Here's one example.

> Decide what you're passionate about, learn the area, and then stay there as long as you can. This allows you to build upon a solid foundation that only strengthens over time.

The Colorado River Redirection Project

This water redirection project presented to me sounded feasible in all areas I consider when I debate investing in a deal. It is a more or less simple process—build a pipeline from the Colorado River in Oregon to California, then place spigots where you want to drop the water off: San Francisco, Los Angeles, or San Diego. Sounds doable enough. But it's a huge project. HUGE. It would take billions of dollars to complete, and the regulations are complicated. Politics from two states were involved, and most importantly, I don't know anything about water redirection. It was completely out of my wheelhouse.

The state of Oregon was very interested to sell its Colorado River water rights to California, as they needed the money. California was interested in buying the water from Oregon, as they are being fined millions of dollars annually for taking too much water from the river well above their legal allotment.

On top of that, I just didn't really feel the pain of the problem. My water bill has been a hundred bucks a month for years. Is this really a problem? Are people in California dying because they can't get water? Yes, but not in the areas this project was embracing. Would this become a significant problem in ten to fifty years? Maybe, but I don't know that for sure because it's not my area of expertise. I specialize in neurovascular devices. What do I know about redirecting rivers? Nothing. So, I passed on the deal. This is a good example of how familiarity with all aspects of a problem improves your chances for success. Stay in your lane.

Stay Flexible about Your Conclusions

Sometimes the solutions you come up with to solve one problem will wind up solving another problem entirely. Viagra, for example, was invented to treat hypertension. Certain cancer drugs are also an effective treatment for asthma. The microwave was originally invented as a radar device for ships. Bakelite, the world's first synthetic plastic, began as an experiment to create better insulation for wiring, which turned out to be way too expensive to suit its original purpose. So, Bakelite became a material for use in high-end accessories instead.

Of course, there are products that have unintended and harmful consequences. Agent Orange, for example, was initially created to speed up the growth of soybean plants, and TNT was originally developed as a yellow dye. Just like when you have kids, when you are bringing something entirely new into the world, you have to remain open to the unexpected and learn to be agile enough to see things differently from the way you saw them when you first went to the drawing board. And you need to be prepared to cope with unintended consequences.

Do a Voice-of-Customer Review

An important step in verifying whether or not you have a problem and a solution that is worth your time, energy, and resources is to do a voice-of-customer (VOC) review, asking the people in your field and your target customers what they think of the problem you have identified and your solution. You can't develop any product in a vacuum. Reach out to the people you are innovating for. If you bring a product to market that few people need or want, then obviously you won't have a successful venture. A VOC is a way to ask many trusted people at the beginning of the process if you have identified a problem with a pain point and a solution that your market can embrace.

For example, physicians identified a problem in my field that they wanted to address, and they came to me to develop the technology. The issue was something called a chronic subdural hematoma. It's a complicated word, but essentially this is a condition that can result when someone, particularly an elderly person, falls and bumps their head. And just like when you hit your arm or your fall on your butt, a hematoma can develop under your skin, or dura, and the ruptured capillaries and blood vessels start to ooze blood and pool. If it doesn't heal and continues to ooze, then that's a chronic subdural hematoma.

Some doctors whom I have done business with pointed out to me that as a subdural hematoma oozes blood over time, it puts pressure on the part of the brain where the hematoma is located—for instance, the controlling thought process center, or the memory center, or the area of your brain that controls motor function. This is a real problem, because what starts as a simple bruise can manifest as dementia when in fact that isn't what's happening at all, and unlike dementia, it should be able to be treated. If you can alleviate the pressure from the hematoma, then normal cognition can be restored.

Presently, these types of occurrences are treated with invasive surgery. A burr hole is drilled through the skin and skull. Then the clot or oozing blood is sucked out, alleviating the pressure to the surrounding area. The problem with this procedure is that (a) obviously, it's invasive, and (b) it's not entirely durable. In other words, it may only temporarily solve the issue caused by the damaged vessels. The underlying problem isn't being solved entirely. The ruptured small vessels causing the oozing are still damaged and may cause more oozing and a subsequent, or chronic, reoccurrence of what the physician just treated. This may reoccur in weeks or months or a year later, and you really don't want to drill a burr hole into somebody's head twice if at all possible.

A group of physicians approached us and asked if we'd be interested in taking a different approach to help them solve this issue by embolizing the vessels proximal to the oozing and permanently treating the damaged microvessels. In this way, we would be able to put an end to any future oozing or subsequent impairment to the patient without ever drilling a single hole in the patient's head, which is always a good thing.

In order to validate whether or not our audience, which were other doctors, recognized this problem and wanted a solution, we did a VOC. We were trying to verify if this problem was widely perceived and if the solution was going to be accepted. If the VOC showed that it was accepted, then we could go ahead and do a clinical trial and eventually a limited market release prior to a full commercial launch.

This is an example of how we would use a VOC in our business to check off all the boxes in the problems and solutions quizzes that I use to validate every process. Regardless of the sector you are in, whether you are inventing a new medical device or a new Frisbee, every invention should include a VOC review that verifies that you have targeted a recognized problem and come up with a solution that will be accepted. You can also learn other valuable information, like the size of your market, what is already out there that works, and what design specifications you will need to respond to your customers' feedback and concerns.

A VOC is going to lead directly into your marketing specifications. The marketing specification is a document that says, "Okay, we have a verified market opportunity, this is the size of it, and here is what our customers have told us in the VOC." It will also communicate the features and benefits we will need to include based on the voice of our customers.

This information also flows into a design specification document, which is a guide that outlines what the engineers are going to design, test, build, and gain regulatory approval for. Then, when it comes time to do a limited market release (LMR), you want the responses that come back in this phase to satisfy the questions and demands that were expressed in the VOC. If your VOC and LMR match up, then you have verification that you actually did solve the problem. You met your design specs, you met your market specs, and your product is functioning as you had intended when you created the original plan. This is giving you a bird's-eye view and sounds simple. It's not. This will be your road map throughout the process, and you'll refer back to it often. It's also the standard by which you will measure success. So think this exercise through carefully, and be sure to create a map that gets you to where you want to go—and gets you there in one piece.

Don't Stop 'Til You Get Enough

Any solution will require a number of iterations in order to realize your envisioned result, and you have to be willing to stay flexible and go back to the drawing board a number of times until you get it just right. It is often said that books are not written; they are rewritten. The same is true of inventions. They are reimagined many times before and after they go to market.

With the stent we developed at MindFrame, we had envisioned that the device would function differently than it actually did once it was in the body. We thought flowing blood would be enough to unbind the clot and we wouldn't necessarily need to remove anything from the artery. We knew physicians were stenting clots in place and that was providing benefit—maybe not 100 percent, but it was having some success. We thought we would function along those lines and just help the clot to dissolve itself.

Our idea was to place our device and allow it to open the artery by pushing the clot against the artery wall and using the stent to allow blood to flow around it. Blood is a natural lytic and will dissolve a clot. Blood contains tissue plasminogen activator (t-PA), which is supposed to accomplish this. Tissue plasminogen activator is a protein involved in the breakdown of blood clots. As an enzyme, it catalyzes the conversion of plasminogen to plasmin, the major enzyme responsible for clot breakdown. There's also a synthetic form that can be used intravenously to speed up the natural process in patients who have experienced a blood clot primarily in the brain, as I described earlier.

So, we thought, "Let's create flow in the vessel and have the blood dissolve the clot naturally. Then we will remove the temporary stent. Bing-bang-boom, and we're done, right?"

That's what we thought. But unfortunately, things didn't quite work out the way we had planned. I recall in the first thirty patients we tested the device on, it worked the way we had projected only twice. Those are not good numbers. Clearly, we had to adjust our technology and develop a device that would capture the clot in the stent structure and then pull out the clot along with the apparatus. We were able to accomplish this and remove the clot, so that in a subsequent angiogram, the physician would see that the artery was entirely clear. The clot or occlusion was removed, and the patient had a normal artery again. Bingo! We had managed to pivot and save the day, along with a lot of stroke patients.

THE STRANGE HISTORY OF THE MICROWAVE

The microwave oven that we have all come to depend upon to pop our popcorn or reheat lunch was actually invented as a radar transmitter during

World War II. Yes, it's true—the microwave oven was a complete accident that was meant to track enemy communications. Percy Spencer, who was an engineer working for Raytheon, was developing technology to create microwave radio transmitters. One day, while testing his machine, he noticed that a candy bar he had in his pocket had started to melt. The microwaves he was testing were cooking his Clark Bar.

This gave the perceptive and flexible Percy an idea, and he began experimenting on concentrating his microwaves to see if they could heat food. The first food he tried to cook was popcorn, which worked like a charm, beginning the timeless and passionate romance between popcorn and the microwave oven.

What Percy discovered was that microwaves are absorbed by water, fats, and sugars and are immediately converted into heat, which cooked food quickly and evenly. Raytheon patented Percy's invention in 1945, and in 1947 they marketed the very first commercial microwave oven, which they called the Radar Range.

Posting Your Benchmarks

Here are a couple of short lists of the benchmarks that every problem and every solution should hit if you are going to successfully bring a product to market. Post these in your innovation team's locker room for easy reference to keep everybody on point.

The Five Benchmarks of Every Good Problem

1. The action or solution currently in use isn't easy or effective in solving the problem.

2. The current solution for the problem breaks regularly under use, is not durable, or won't last.

3. The current solution to the problem causes downstream unintended damage to something or someone else.

4. There is nothing currently available to solve the problem, or what there is can't achieve the results you are looking for.

5. Current solutions to the problem are too expensive to access.

The Five Benchmarks of Every Good Solution

1. The solution involves tasks that are simple and easy to accomplish.

2. The solution is less costly or less expensive than the current one.

3. The solution is more durable or reliable than the current one.

4. The solution saves lives, prolongs life, or enhances life.

5. The solution allows you to do more in the same amount of time through automation and reduces the potential for human error.

CHAPTER 2

OWN YOUR SOLUTION

When you win, say nothing; when you lose, say less.

—Wayne Gretzky

I loved playing hockey when I was a kid, and I was pretty good at it too—and I knew it. But when I started eighth grade, things got real. I was going to be entering high school the following year, and I knew if I wanted to play high school hockey at St. Bernard's, then I was going to have to up my game. This meant taking a look at my features and benefits realistically, figuring out my differentiators, and assessing where I stood the best chance of making the team.

The challenge was, as usual, that I was a generalist. I was good at a lot of things but not stand-out stellar at any of them. I knew I could skate with anybody, but I wasn't a star. I wasn't the fastest; I wasn't the best goal scorer or the best passer. But I was solid, I was reliable, and my legs were big, so nobody was going to knock me down. I was a grinder, I thought. That's my ticket. The only thing was, I had some pretty fierce competition in my space. There were more than a few grinders in my neighborhood. I grew up in a French-Italian town full of big, strong kids like me, so I had to really shine if I wanted to stand out as a grinder. I was going to have to become the best solution for my high school hockey team overall as well as in my position. To do that, I needed to build up my strength and endurance so there would be no question about who was the best choice.

That summer, I came up with my own proprietary eighth-grade grinder training program, which I was sure would give me the edge on the competition that I needed.

I bought a pair of Rollerblades, and I'd skate up hills all day long wearing ankle weights. This was my anchor concept. The weights were a stroke of genius. And then, after skating up hills all day long, I practiced my shots in my backyard after dinner, and I'd wear wrist weights to build up my forearms.

The competition started to notice. My friends were like, "Dave, what are you doin' man? You're huge!" But I never told anybody my trade secret except my dad, and he kept my secret. This was big of him. I had shot pucks that had broken out more windows on the side of our house than you could throw a hockey stick at. And I might have done that at some point too. I wrecked the joint, but that fall, I became the only freshman to make the varsity team that year. If I hadn't taken the steps I needed to take to develop a novel method and keep my trade secrets a secret, who knows whose name would have wound up on that starting roster.

You Are Not Alone, and That Should Not Reassure You

Big ideas happen more like waves on a shoreline than like a light bulb switching on. The process is more gradual, and it takes a multitude of little waves just like yours to move a tide. All this to say, if you've come up with a great idea to solve a common problem, you can bet that somebody else out there has probably thought of the same thing and is probably working to develop it right now, just like you are. In the world of innovation, you are not alone. Things move in waves, not rivulets. Don't forget that.

Particularly when you're innovating in a specific sector, like I am in interventional neurovascular radiology, the science gets to a certain level of development, and then everybody starts experimenting with the next generation of technology. Plus, we all know each other. We all go to the same conferences every year and hear the same speakers. We all read the same publications. So when you really enter a space and begin innovating, it's only a matter of time— and usually not very much time—before your colleagues arrive at a similar conclusion. *But only one person can own the solution, and that is the person who owns the patent.*

Here's a story from my case files about the time I learned firsthand about the importance of speed to patent. Because once a seed is planted, it's going to grow.

> Only one person can own the solution, and that is the person who owns the patent.

The Covidien Treaty

When we filed for a patent at MindFrame in the US, we decided to file in other countries outside the US that were part of the Patent Cooperation Treaty (PCT) first. We made this decision because you should patent in every country you want to sell your product in anyway, and the process moves much more quickly in certain countries like the UK, France, or Germany. The United States takes forever. Plus, once you have a patent in one PCT country, your US patent filings get bumped to the front of the line.

The UK wasn't a huge market for us, but they were part of the PCT, the examiners over there are all PhDs and ubersmart, and the process was superfast. We had a patent issued in the UK within nine months. Then we got bumped to the front of the line with our US patent application, and we were issued a patent on our device in the fall of 2011, well ahead of all our competitors—which, as it turned out, was a very good thing.

Around that same time, another company, Covidien, developed a device called Solitaire, which was a device designed similar to ours, but was actually designed and indicated to treat brain aneurysms, although the physicians weren't using it to treat brain aneurysms. However, if you used their device in the novel way that we intended, then their device could do the same thing that ours did and successfully treat ischemic stroke patients.

Covidien was caught off guard by this. They had never realized their device could be used in that fashion. They may have thought it was too dangerous. We and countless physicians had just proven that it wasn't by using their device technically off-label to successfully treat ischemic stroke with large-vessel occlusions. And patients got better. Nobody died. Covidien already had the technology that would accomplish the same goal that we were after; they just hadn't filed a patent for use in ischmic strokes.

I attended a meeting in France on January 13, 2009. The meeting is called ABCWIN, which is a working group in interventional neuroradiology, and it takes place in Val d'Isère, France. This is something I love about my field—I can go places thousands of miles away from Orange County, California, where I live, and I have friends. I feel right at home because I've been there many times before. I have made friends all over the world who graciously welcome and entertain me wherever I go. This is a huge perk of my line of work for me. I really love that part of my life. The ABCWIN conference is one such opportunity to see old friends and meet new ones. I have been attending without missing it once since 1997, and I still attend to this day and will continue to, however that becomes possible in our postpandemic future.

That particular year, Professor Thomas Liebig was presenting a case series called "Endovascular treatment of acute thromboembolic occlusion with a self-expanding fully retrievable stent." Sounds like a real toe-tapper, right? Well, for me and a lot of folks like me, it was. Dr. Liebig was a hot ticket.

The doctor presented a series of twenty-six patients whom he had treated using the Covidien Solitaire aneurysm device to remove blood clots in patients with large-vessel occlusions who were experiencing acute ischemic stroke (AIS). The audience was baffled. I took pictures of each slide he presented. I was stunned, because I could not believe what I was watching and witnessing. He was treating patients with a device we were developing to do the same thing. We thought we were the only ones. I felt a sinking feeling in my stomach.

Had Covidien filed patents before us? But that couldn't be. We filed our provisional patent application in the spring of 2007. We converted that into a patent application in the spring of 2008. Our patent had not yet published since it hadn't been eighteen months since we filed a patent application. Nobody had seen our idea yet, and nothing had come up in our search of the art prior to patent. They couldn't have patented before us. Could they? I had no idea. I had to wait and read the *PCT Gazette* the following month to see if their patent application would publish before ours.

Nineteen months after Dr. Liebig gave his presentation in France, their patent application published. Twenty months later, Concentric Medical's patent application published. They were way behind us, and their application was preceded by a presentation in a public forum by Dr. Liebig on January 13, 2009, which would make it general use and not patentable (more on that dilemma later). I knew then that we had them. It was only a matter of time before our patent would issue. But would we get the right claims?

After their presentation at ABCWIN in 2009, though, Covidien did start to market their product as a device that functioned like our product. Our patent had not issued yet, so they were free to do whatever they wanted to do. I don't believe they knew we had even filed. They were approved for general neurovascular application in Europe under CE Marking, so even though we were the pioneers in the space, they'd be first to market while we waited for our patent to issue. Covidien was a huge company with a big experienced sales and marketing machine. They could squash us in the marketplace if our patent wasn't approved.

Fortunately, because of our filing strategy, our patent was approved in 2011, and Covidien realized they were infringing on our patent. Our chairman, who was friendly with the president of Covidien at that time, said, "Come on, let's make a deal," and eight or so months later, they acquired us for $75 million cash.

That was a happy ending, but it was only a happy ending because we had the patent FIRST. If our patent had been held up or denied, Covidien would have annihilated us, because as a big company, they were in a position to beat us to the punch and implement our solution.

Just What Is a Patent, Anyway?

A patent is the grant of a property right to the inventor, issued by the United States Patent Office as well as by similar agencies in other countries. The right conferred by the US patent grant is, in the language of the statute and of the grant itself, "the right to exclude others from making, using, offering for sale, or selling" the invention in the US or "importing" the invention into the US.

In order to obtain a patent in the United States, you have to file a US patent application. You can either file a design patent application (which covers the way something looks but not the way it functions), a plant patent application (to cover asexually reproduced plants), or a utility patent application. The utility patent application covers what most people refer to as an invention, namely devices, methods, compounds, and software.

In the US, anything useful under the sun made by humans can be patented, including business schemes, pharmaceutical compounds, and computer programs. Discoveries or scientific theories are not "made by man" so can't be patented. You couldn't have patented gravity if you were the first to understand it, and you can't patent a new type of insect that you've discovered, although you can patent a genetically modified animal or plant. There is a lot that can't be patented. Here are some examples.

According to the Patents Act, an invention must constitute more than the following elements in order to be approved for a patent:

1. A discovery, scientific theory, or mathematical method

2. An aesthetic creation

3. A scheme, rule, or method for performing a mental act, playing a game, or doing business, or a computer program

4. A presentation of information

5. A procedure for surgical or therapeutic treatment or diagnosis to be practiced on humans or animals

In order for an invention to be patented, it must have a useful purpose that is novel and nonobvious. You can get a utility patent for new compositions, production processes, machines, and tools. You can also patent an upgrade to or improvement upon an existing innovation. Here is a punch list of elements that your invention MUST include in order to be approved for a patent:

1. In order for your invention to qualify for patent eligibility, it must cover subject matter that Congress has defined as patentable. The United States Patent and Trademark Office (USPTO) defines patentable subject matter as any "new and useful" process, machine, manufacture, or composition of matter. Machines or processes are patentable subject matter, but the laws of nature are not. So, you can patent a machine that bottles spring water, but you can't get a patent on the spring.

2. The invention must have a "utility" or, in other words, be useful.

3. The invention must be "novel," or new.

4. The invention must be "nonobvious," meaning its use or function can't be something that is simply the next logical step of an already patented invention. Much of the argument between the USPTO and patent applicants revolves around the issue of nonobviousness.

5. The invention must not have been "disclosed" to the public prior to the application for the patent. For example, if you've written an article describing the invention before you apply for the patent, the USPTO may deny the application because you've already disclosed the patent and therefore it's public knowledge. Remember the ev3 (Covidien) patent application that published nineteen months after Dr. Liebig's public presentation?

Lawyer, Lawyer, Lawyer

The first word in innovation is *Eureka!* The second, third, and fourth words are *lawyer, lawyer, lawyer*. Over the years I have had the opportunity to innovate and work with many inventors. Many of them are first-time inventors, or engineers or doctors who are for the first time attempting to protect their own invention. I have found that those who are serious about their invention are also capable of meaningfully participating in the preparation of their own patent applications. But even these highly capable applicants should not attempt to file their patents by themselves unless they are also a patent attorney.

These folks are motivated, but they simply do not know what to do or exactly how to do it, and they are afraid to mess things up by trying to do something themselves that is over their head—and they are right. This isn't because they're not smart enough; it's because patent law is dense, unapproachable, and very complicated. The rules are complex and unnecessarily so in many, if not most, situations, and they change. It almost seems that some rules have been created solely for the purpose of tripping up the very people who need access to the system.

I don't believe that is how the rules were conceived, but it certainly is how they have evolved over time. That being the case, should inventors be going solo and trying to protect their own inventions? NO! That would be similar to trying to remove your own appendix. If you have any chance at all of getting to a hospital, you should do that instead of trying to operate on yourself. You would do anything possible to get yourself to a surgeon, and you should feel the same way about a patent law. It's as complicated as medicine. You should not try to do it yourself. It's not safe.

When you are an inventor or entrepreneur and you have a dream, there are a lot of things you have to do, and it can seem like there's never enough money to do everything you should do. Then, you confront the idea that you have to hire professionals in order to really do things the right way. This is the inventor's and entrepreneur's dilemma, but you have to rise to the challenge. A patent is going to frame out your future and your future potential, and a lawyer can help you understand how to make sure that you have a framework that will allow you to grow and prosper.

In most inventors I meet—and as the head of a venture studio I meet them every day—I find that beginners have the idea that the assistance of a patent attorney is maybe one corner they can cut. This is an unfortunate assumption on their part, because those who represent themselves ALWAYS obtain rights that are narrower than they otherwise could have been, sometimes so narrow that they are completely worthless from a commercialization perspective.

Inventors who file their own nonprovisional patent applications and subsequent patent application frequently get a first office action (FOA) that rejects all the patent claims they filed. An FOA that rejects all claims is not uncommon, but most applications filed by individuals have little or no useful discussion of the invention. That means there will be little or nothing anyone can do to help them ever achieve a patent. Sometimes inventors understand that they made a critical error, but usually they blame the patent examiner or a rigged patent system. The truth is they didn't file an application that was legally adequate.

Cutting corners by trying to do your filings by yourself usually isn't the most economical approach in the end. Simply stated, a bad patent application results in either an extremely narrow patent or no patent at all. All that time, money, and energy have been wasted. The new inventor, who is unfortunately the norm for those who represent themselves, may well have had an invention that could be protected, but through a faulty initial patent application, they will now likely never receive a patent on their invention. It may not seem fair, but we do not have a "fairness system"; we have a legal system. And when dealing with patent procurement, the law is what the law is. There are many pitfalls waiting to trip up the unwary.

Words of Warning for Patent Filers Flying Solo

First and foremost, those who are going it alone need to remember that the job of the patent examiner is to examine what you present, not to help you obtain the broadest protection possible or any protection period that would be useful in a commercial sense. Never forget that this is an adversarial process. On top of that, many things can be hopelessly compromised at the outset of the patent application process that can make it impossible to obtain a patent of any kind or at the very least force you to file another patent application to correct the deficiencies present.

If you do have to refile, you will then be giving up your original filing date, which can prove fatal. Therefore, inventors need to appreciate that the first filing is an all-important, critical filing that absolutely must describe the invention and all of the variations in as much detail as possible. You need to describe everything from the broad general invention to the most specific version of the invention and everything in between.

One of the most common mistakes people make at the beginning is being unable to describe what their invention is and what patentable novel feature or unique contribution their invention is making to the relevant technology field. Another common mistake is to describe everything with equal importance in a claim when there are almost certainly features that deserve greater attention because that is where the patentable invention is anchored.

I think back on the time when I was young and I told my patent attorney, "I have never seen anything like my invention on the market, so I know there is nothing that could stand in my way of obtaining a patent." Those were more innocent times for me, ones that quickly evaporated as soon as I learned how much I didn't know, which always teaches you humility. Or at least it should.

There are a lot of reasons why something identical to or similar to your product may have already been patented or even just had an application filed, which is enough to render your product public knowledge and therefore not eligible for a patent. Many times, an independent inventor will obtain a patent and then run out of money or lose interest, and therefore the product never gets developed commercially. This happened to a company I cofounded called Neurvana.

At Neurvana we along with a couple of physicians coinvented a device and method to treat vasospasm. Vasospasm is a vascular issue and a by-product of a subarachnoid hemorrhage (SAH) in the brain causing narrowing of the arteries in the affected region. This can occur in up to 30 percent of patients with SAH. Some physicians came to us with a problem and for the initiation of a solution. They had already used a device off-label to prove the concept. We filed patents. The patent was subsequently issued, but the company failed to commercialize the product technology because they went out of business. Sometimes people even do this on purpose just to wait until somebody else comes up with the invention and develops it, then suddenly they remember their patent.

Due to the laws of nature and the reality that there is only a finite number of solutions to any particular problem, every generation invents or reinvents many of the same things. This is why doing a patent search is absolutely essential. Until you understand what is known, you have absolutely no way of knowing whether a patent is likely to be obtained.

I guarantee a patent search will uncover inventions you did not know were out there. With over ten million utility patents having been granted in the US and millions of published patent applications in the US alone, something can always be found that at least relates in some ways to your invention. You are always better off knowing about those related inventions. This allows you to determine whether moving forward makes sense, and it also allows for a patent application to be written to accentuate the positive, and likely patentable, aspects of an invention.

The moral of the story is to be careful. Getting help from a trained patent professional is the best and safest way to proceed. If you do not have the funds available to seek competent professional advice, you should really ask yourself whether you should be pursuing the patent path at all.

We all have dreams, and sometimes we take risks, but the invention-to-patent-to-commercial-success path can be long, time consuming, and expensive. I dream big myself, so I am not about to tell others not to follow their dreams and believe in themselves and their inventions, but it is best to go into the process understanding what lies ahead and pursue moving forward in a financially responsible way.

THE STRANGE HISTORY OF THE TELEPHONE

Many of the world's most groundbreaking inventions were being developed simultaneously in different places and by different people around the globe. And many brilliant inventors are forgotten by history because somebody else got the patent before they did or filed a patent that was more accurately worded and allowed for growth into new and unforeseen purposes.

The only reason that we remember Alexander Graham Bell as the inventor of the telephone is because his competitor worded his patent application incorrectly. There were at least four inventors of the airplane, and a guy named Tesla invented electricity long before Thomas Edison lit his first bulb. But Thomas Edison paired up with General Electric, who got the patent first.

As all of us know, the telephone was invented by Alexander Graham Bell. A teacher for the deaf, Bell wanted to invent a way to telegraphically transmit speech in order to better communicate with his students and with his wife, Mabel, who was deaf. Mabel's father, Gardiner Hubbard, and a group of his investors asked Bell to instead help perfect the harmonic telegraph, which would ultimately become the telephone. Hubbard agreed to let Bell work on both, and ultimately, as well all know, the telephone won out and was patented by Alexander Graham Bell in 1876.

Years before, however, two other people had also invented the telephone. Antonio Meucci came up with the idea of a talking telephone in 1849 and filed a patent on his design in 1871. Unfortunately, he couldn't afford to renew his caveat, and he lost control of his innovation and his place in history. Elisha Gray, a professor at Oberlin, also invented the telephone and applied for a patent on the very same day that Alexander Graham Bell did. Bell's lawyer just got to the office first. Alexander Graham Bell's application was number thirty-nine that day, and Gray's was number forty. And the rest is history.

Good Patents Make Good Partners

Until you are able to file at least a provisional patent on your invention, try to limit how many people you talk to about your solution. Don't talk to people or investors; don't post it on Facebook or Instagram or hashtag it on Twitter. Mum is the word until you're protected. I've run several development groups at medical device companies and now in my venture studio; whenever anybody comes to me with an idea, I tell them not to say another word to me until they have filed at least a provisional application.

I'll give you an example of what I mean. A good friend of mine introduced me to his customer, Dr. Sheth, who is a brilliant interventional neurologist. He wanted to meet to ask for help one day because he had a great idea for a new device.

"Dave, I've been meaning to talk to you," he said, and his eyes got really big, so I knew he wanted to tell me something really important to him. "I know that you're involved in developing new medical technology and bringing new products to market, so you are just the guy I need to talk to. I've come up with an idea for a device that can first record electrical brain activity and then be implanted in the brains of epilepsy patients or people with seizures. It's sort of like a pacemaker for the brain, and it can monitor, locate, predict, and prevent seizures in real time."

Now, Dr. Sheth was really saying a mouthful. There was nothing like this currently in the market. The closest thing you could get to locating a seizure was by having the patient wear this crazy-looking skullcap, which could only be done in a doctor's office, or getting a service dog that recognizes the symptoms. This was a real breakthrough, and I knew it. We could make a huge difference in people's lives with this invention and, obviously, make a fortune if this solution really worked. So, while it might seem counterintuitive, the first thing I did was stop him from telling me anything else.

"Listen, Dr. Sheth, you have identified an important problem, and you clearly have an innovative and potentially viable solution. So now I am going to do something that's going to make you think that I'm pushing back, but I want you to know that I'm not; I'm just protecting your rights." He listened carefully and then tried to say something else, and I put my hand up.

"Before you say anything else to me at all, I'm going to introduce you to my patent attorney, Peter," I explained, feeling a little like Moses on the mountaintop. I was trying to teach him the first commandment of innovation: thou shalt get a patent attorney.

"I want you to talk to Peter—download this for him; it will all be kept strictly confidential—and then he will help you with a provisional patent application. Once you've done that, I will introduce you to my business attorney, Goody, and you're going to form an LLC and assign your provisional patent application to the LLC. Then you can come back and tell me all about it, and maybe we'll do business."

"I can't afford to do all that," Dr. Sheth said, looking overwhelmed and little bit offended. I mean, after all, that's why he was coming to me for free advice—because it was free.

"If you are right about this device, you can't afford not to do all of that," I told him. "You've been to the medical conferences; you know how it is. Doctors think they're the new entrepreneurial class. You'll get ripped off in a second, and they won't even realize that they're doing it. They'll think they thought up the whole thing on their own, and the courts will agree with them unless you have filed for ownership of this thing."

And then I gave him Peter's number and let a billion dollars of potential revenue walk right out of my office. Why did I do that? Because ideas are cheap. Everybody has them. And an invention without a patent is nothing more than an idea that nobody has invested any time or resources into yet and is therefore worthless. This is the reason that although I'm in the business of helping inventors get their products to market, I don't talk to anybody unless they are serious enough to file for a provisional patent in advance, no matter how good their idea is.

If you can't afford an attorney, there are resources available to help you file a pro bono patent. Call your local law schools or the bar association, because often there are pro bono legal services available to innovators and artists who can't afford an attorney. You can also call USPTO for guidance, and there are a lot of helpful resources for beginning innovators on the United States Patent Office website.

QUESTIONS TO ASK YOUR PATENT ATTORNEY

Here are questions that you should ask prospective patent attorneys in order to decide whether or not you are a good match.

1. What is your technical background? A good patent attorney for you should have experience in your industry and be familiar with your particular technology.

2. How many patents have you drafted, issued, and enforced? You want a patent attorney with some respectable experience executing each.

3. Can you show me your last six issued patents and explain the invention to me?

4. Can you explain the circumstances surrounding your last six issued patents, and what makes your outcome important?

5. Have you ever had any patent applications ultimately not be granted a patent, and why?

Do Your Research in Advance

You can minimize high-rate billable hours by doing as much research as possible on your own to verify the eligibility of your patent and assembling all relevant documentation and material so you don't wind up paying a lawyer hundreds and thousands of dollars just to go looking for things and organizing your stuff. Take the time to under-stand your invention, where it fits into the marketplace, and what you foresee about its viability and usefulness beforehand. Understanding these fundamental elements of your new invention will require research on your part, but it's time well invested. Here are some questions you should be answering for yourself:

1. Who are your competitors?

2. What are your competitors' strengths and weaknesses?

3. Which industry do you fit into, and is it a good fit?

4. Are there related industries where you also fit?

5. Who is the target market for this product?

6. How would you sell your product to them?

7. Whom would you target for first-round investment?

8. Whom might you put on your board?

9. Whom might you put on your team to help you develop and/or launch this product?

It's also a great idea to do your own search of prior patents, or what we call "prior art." The best way I know of, which I have been doing myself since the 1990s, is to go to www.uspto.gov. This is the US patent website. You can search keywords, inventor names, assignee names, and so on. And the best part is, this search service is completely free.

Why Patent Applications Fail

Patents get rejected most commonly due to obviousness and lack of novelty. Examiners don't like it when what you are claiming is not clearly unique and new. Another reason for rejection is when the examiner sees

> Take the time to understand your invention, where it fits into the marketplace, and what you foresee about its viability and usefulness beforehand.

that the idea you are claiming to have invented already exists in the prior art or is already patented. Finally, applications are commonly rejected because what you are claiming is so obvious to any skilled person that it just isn't novel enough. This comes down to the examiner taking one or two claims from another patent and combining them to come up with your idea or solution. This renders an idea so obvious that it is not patentable.

It isn't unusual to get a FOA rejection as a new inventor, so don't panic. You will get another shot at it, but maybe not more than one. A second or subsequent action from a patent examiner on the merits shall be final, except in certain very specific circumstances.

When information is submitted in an information disclosure statement, the examiner may use the information submitted and make the next office action final whether or not the claims have been amended. This means that generally speaking the second action you receive from a patent examiner is going to be a final rejection.

Once you receive a final rejection, you are quite limited with respect to what you can do as a matter of right. This matters, because at this point you almost always want one more amendment opportunity, which may not be available. Once a proper final rejection has been entered, there is no longer any right to unrestricted further prosecution.

Getting Down to Business

The next crucial step in solidifying ownership of your solution is to form a business entity and then assign your provisional patent application to that business. Why should you create a business right away? Because that is what innovators who think they have an idea that is going to make money do, and they'd like to hold on to it.

Being a business allows you to do a number of things that are central to future success, potential liability, and profits. In addition, the type of business entity you choose to form can radically impact how you operate in the future, how you can raise money and invite partners, and so on. The regulations governing each option are as complex as patent laws, so seek professional counsel and find a good business attorney. This is the second strategic player that you will add to your core innovation team. Your business attorney should be someone you trust, who is able to understand everything that you want now or may want in the future, and who can help you craft a perfect scenario to help you achieve those goals.

QUESTIONS TO ASK YOURSELF BEFORE YOU SEE A BUSINESS ATTORNEY

- What is the nature of your business?

- Are you providing a service or product?

- Will you have partners?

- Are you raising money either through an equity financing, convertible note, or line of credit?

- If you are raising money, are you raising from US investors or international?

- Are you raising from venture capitalists or family offices?

- Will you eventually sell the company or product owned by the company?

- What state will you operate in?

All of these questions will impact what kind of a business entity you select, and some have restrictions that will not allow you to do things you may want to do down the line. So it's important to understand the nature of each, what they offer you, and how they might restrict you.

The Three Basic Business Entities for Entrepreneurs

- **LLC: A Limited Liability Company**

  An LLC is a corporate structure that is a hybrid containing elements of a partnership or sole proprietorship and elements of a corporation. LLCs are known as pass-through entities in that all debts and liabilities are passed through the organization so the owners are not personally liable. An LLC is arguably the simplest and least-regulated alternative and offers a great deal of protection. There are some downsides. One is that venture capital firms can't invest in pass-through entities like LLCs, so if you want to raise venture capital or become a publicly traded company, you should not select this route.

- **S Corp: A Subchapter S Corporation**

  An S Corp gives a business owner the benefits of incorporation while being taxed as a partnership. Subchapter S corporations are also pass-through organizations in that liabilities and losses can pass through them rather than the owners. Additionally, income can be passed directly to shareholders. S Corps do not pay taxes at the federal level; rather, the shareholders pay taxes on the income they receive from the corporation. Saving money on corporate taxes is beneficial, especially when you are just starting a business. S Corps do establish a certain level of credibility for new business; however, they are far more complicated regulation-wise than LLCs. Additionally, because they are also pass-through entities like LLCs, you cannot raise foreign money or attract foreign investors.

- **C Corp**

Most major companies and many small-to-medium businesses in the United Sates are C Corps. C Corps enjoy the
same benefits as S Corps; however, C Corps are taxed as
business entities at the federal level. This means that C Corps
pay corporate taxes on earnings before distributing dividends
to shareholders, who are also taxed on those dividends. In
order to avoid double taxation, many shareholders reinvest
their profits back into the business. C Corps can also have
their stock qualify as QSBS (qualified small business stock)
under the IRS code 1202. This is complicated but valuable to
specific investors. If you've held a stock qualifying as QSBS
for at least five years when it's sold (more on this point later),
a portion of your gain, or in some cases all of your gain, can
be excluded from federal tax. There are lots of regulations
and requirements involved with C Corps, including voting
records, bylaws, annual reports, and financial disclosure
reports; however, you can raise venture capital and foreign
money. So, if you are thinking really, really big, you will have
your eye on a C Corp, but rarely if ever do I recommend that
businesses start there unless investors request it.

BUILD IT, AND THEY WILL COME

*Just because something doesn't do what you
planned it to do doesn't mean it's useless.*

—Thomas A. Edison

Now that you have a documented solution that you provisionally own, the next step is to build your first prototype and start to play around with it. You will learn a lot about your solution during this phase, because you are now actually going to reduce an idea into practice. In this phase, your idea becomes a thing that exists in the world, something that did not exist before, and this can be an exciting but sometimes terrifying process.

Once you convert a provisional patent application to an actual patent application, you will eventually have to defend that application when you get an office action from the PTO. A lot of people get their claims rejected by the examiner. This can be particularly true when you're a beginner. I've had so many patents rejected in my tenure that I can't even remember them all. You can get around this. I did. So, don't despair and just give up.

Even with a good lawyer, the patent process is trying, but if you persist, you can change the expert's mind. You just have to get strategic. You don't have unlimited rounds to convince the examiners before they issue a final rejection. You get three chances to convince them of the novelty and purpose of your invention, and then you're officially toast.

I like to go in and physically demonstrate my solution for the examiners rather than try to explain things on paper that they didn't understand the first time around. This isn't always needed or even possible. However, I like to get face to face with the examiner and help him or her to really understand what I'm talking about. I can only do this, though, if I have a working prototype. So as soon as I file a provisional, the pressure is on to come up with a working example that I can use to defend my position.

> Even with a good lawyer, the patent process is trying, but if you persist, you can change the expert's mind. You just have to get strategic.

Another compelling reason to move efficiently into the prototype phase is that you are not the only one who is developing solutions in your category. Other people are trying to figure out the big problems, too, and so from the time you file your provisional application, you have only twelve months to convert the provisional into a real patent application, or it expires and you lose your original priority date. Remember that the "first to file" date can be a game changer.

The good news is that it stays confidential, but the bad news is that you lose your priority date. If you do convert to a nonprovisional patent application, then in eighteen months the patent application publishes, and people like me who read the *PCT Gazette* on the first Tuesday of every month will know what you're up to. Sounds like a lot of time, but time goes by quickly when you're having fun.

This leaves you only one to two and a half short years to conceal your idea, build your prototype, test it, retest it, and convert it into a product. That's not very much time, and there are ways to work within the system to expand that window, but essentially once you file an application, the heat is on. Once the cat is out of the bag, your primary advantage will be speed to market, and that will largely depend upon how efficiently you can arrive at a viable prototype.

Work Your Network

Depending upon the scale of the solution you are contemplating, the expense of building a prototype can vary. If you require a lot of expensive development and manufacturing capability, it can get very expensive, in which case you will want to bring in technical and investment partners to assist you. Other folks build prototypes all on their own using framing wire, spit, and duct tape.

Regardless of expense or the complexity of your design, there are a lot of issues that arise when you go from concept to a physical object (or art to part), and it's a good idea to have people on board with experience in designing and building solutions like yours. This means getting out and connecting with other people who are thinking about the same problems and solutions that you are. If you don't network and start to hang around with people that have similar interests, you can really slow yourself down at this juncture.

The prototype phase is the time to begin the general conversations that can lead to important alliances down the road. Now is the time to meet the folks you may want to bring onto your team as your product develops. So, get out there and meet the great talent in your space. Join groups and organizations; go anywhere you can to meet the people who are already doing what you do so that you can access their expertise when you need it or even just commiserate with somebody who understands exactly what you're going through.

How to Choose a Development Partner Who Won't Drive You Crazy

In the early concept phase of medical device product development, depending on the solution you've come up with, finding a development partner to reduce your idea to practice can be either very straightforward or very, very challenging. There are many contract manufacturers that can help you. You will be able to find development partners at industry expositions as well as market-specific conferences. I've done it. I've found that this be can applied to most any catheter-based product, and that's pretty specific. But be sure you notice that I said *most any* rather than *all*. Sometimes, solutions involve technology that nobody has thought of yet, in which case you will have to go it alone and improvise until the times and the technology catch up with you.

I encountered this with my team in 2007/2008 at MindFrame. We had a concept that we believed would solve a problem. We had to start, however, with a larger device because the nitinol tubing we wanted to use hadn't been developed yet in the size we needed. It wasn't until two years later that the technology finally became available to make the size we needed to bring our solution to fruition.

At conferences you will find contract manufacturers and commercial companies in your field that will want to either develop your idea for a fee-for-service contract or, in the case of a competitor, will want to codevelop with you to get your commercial rights and acquire your company when the product gets to market. Therefore, you have options—you can go it alone with a contract manufacturer, or partner early with an industry player who believes in you and your solution and is willing to wait.

QUESTIONS TO ASK A POTENTIAL DEVELOPMENT PARTNER

Here are five questions that I ask to evaluate whether or not a development partner is right for my project:

1. Is your solution within your capability and wheelhouse?

2. Have you worked with the materials or technology I need to bring my solution to market in the past? Are you working with these materials currently?

3. What is your budget estimate, and can investors afford you? Or are you too expensive at this juncture?

4. What is the timeline? How long will it take you to make a prototype? Weeks? Years? Does this match with your investor's timeline to go to market?

5. Will you sign a nondisclosure agreement? It's critical to keep your idea confidential despite the fact that you've already filed patents, so make sure you work with a partner who will protect your trade secrets.

Ethical Entrepreneurialism

Perhaps more than in any other field, as an entrepreneur you face a lot of tests of character and moral dilemmas in the course of your day-to-day business. This happens when you're the first to any frontier. There aren't any road signs or maps. So when you come to a crossroads, you have to decide: Which way am I going to go? Who am I? Am I a shortcut or a long-way-home person? What are my values? Do I have to take the high road, or can I live with the low road? What is most important to me? When this moment arrives, as it inevitably will, I've seen people do the right thing, and I've seen people do the wrong thing. This includes myself. You hit a lot of the fundamental ethical questions during the prototype phase.

In my career, I've made some really good decisions, and I've made some pretty bad mistakes. Everybody makes a few, especially at the beginning. Mistakes are good, as long as you learn from them. When you refuse to learn the lessons of your experience, you can get onto thin ice very quickly, and before you even know you're in trouble, the ice breaks.

Theranos–A Modern Greek Tragedy

Theranos is an epic cautionary tale about what can happen to you if you refuse to realize that you have hit a wall.

After the podcast and the documentary and the soon-to-be feature film starring Jennifer Lawrence and eventually, I'm sure, the Broadway musical, just about everybody has heard of Elizabeth Holmes and the company she founded: Theranos. Theranos logged one of the hugest meltdowns in history during its prototype phase and subsequently lost about $900 million of its investors' dollars in the process.

For those of you who have been living under a rock or hiding out in your basement like a mushroom, plugging in stuff to see if it works, let me briefly recap. Elizabeth Holmes—at nineteen years old, no less—set out to change the world. And she really could have. Elizabeth had a great idea, and had she handled the prototype phase more wisely and less emotionally, she might have really achieved watershed innovation. As it was, though, she never made it out of the prototype process. She never achieved an actionable commercially viable product. But that didn't stop her from going to market with it. And that's where the hubris part comes in. She violated the number one golden rule of new product development, particularly in our medical device space—thou shalt not release any product that is not safe and reliable.

Elizabeth, with her original invention, the Edison, set out to disrupt an industry that hadn't really evolved technologically since the 1950s. The medical testing field is ripe for innovation. It's slow, it's expensive, and it's exclusive. But it's all we've got.

Elizabeth's origin story relies on the idea that her mother was deathly afraid of needles, and she saw how this prevented her family from getting the actionable health information they needed. As a result, she lost a beloved uncle while he was still in his prime, and her idea of a testing process with only a finger prick was born. Pretty compelling stuff, right? Well, a lot of really prestigious investors thought so too, like George Schulz, Henry Kissinger, James Mattis, and William Frist, MD, to name just a few.

Elizabeth was right about one thing: everybody hates needles, and diagnostics are just too expensive and inaccessible for many people. Elizabeth's big idea was to solve this problem with a machine that could perform a thousand diagnostic screens from a single nano-container of blood gathered through a finger prick and then deliver results directly to the patient inside of thirty minutes. Think the photo-developing kiosks in CVS, only faster!

Gone were the days of terrifying hypodermics sucking endless vials of blood from wary patients, who leave the labs of America feeling debilitated and disempowered and no better informed about their health than when they walked in. Never again would we have to say goodbye to someone we love too soon because they were too afraid of the test to get the actionable health information they needed. Once Walgreens jumped on board, the Edison and the life-saving information that it provided would become accessible to every American right in their local pharmacy.

The only trouble, of course, was that the Edison didn't work, like, at all. Elizabeth's idea of performing dozens of diagnostics from a single drop of blood was impossible. Any lab tech who worked with blood testing in any capacity knew as much. But Elizabeth and her board didn't accept this. Theranos innovators and engineers slaved endlessly to produce the impossible, but to no avail. And when they told Elizabeth that what she wanted wasn't possible, she fired them and hired new ones.

I myself had a chance to invest in Theranos as part of an angel investor group, and I thought that the Edison, if it was legit, really could change the world. If the Edison had worked, people would have made fortunes. The company was valued at $9 billion the day everything went up in a cloud of smoke. I passed on the deal because as good as it sounded, I know nothing about blood screening, and I have a rule that I just don't ever stray out of my lane. But I did call around and ask a few folks, and everybody who knew anything at all about the subject—and even the phlebotomist at my local lab—knew it couldn't be done. A person's fingertips are dirty. It's a terrible place to gather blood for testing purposes. All kinds of bacteria get involved. And just thinking like an engineer in the space, I felt you couldn't make a box the size of a home printer and expect it to do a thousand screens inside of a half hour. But Elizabeth wasn't listening to the physics.

Elizabeth knew after the first $45 million that it wasn't going to work as planned. She knew because her team had told her so. They had been telling her that they didn't have enough space in the confines of the box to do all of these tests. They needed a box the size of five toaster ovens, not two. But instead of listening to her team, Elizabeth doubled down.

Elizabeth's solution was to rebrand the same machine that didn't work. And so, the Edison became the Minilab, but no matter what she called it, the thing was still an unworkable prototype. You can't solve technical problems with a marketing campaign. You can't just say something and make it so.

To make matters worse, Elizabeth didn't have a single expert in her field on her board, so the board believed her progress reports and didn't push back. So, without anybody there to speak truth effectively to power, Theranos went to market with a product that had never reached commercialization phase. They sold Walgreens, Ralph's, and Safeway a testing device that didn't work, and the rest is a still-evolving catastrophe for everyone involved.

Brick Walls Are There for a Reason

All innovators hit walls in the prototype phase. I have many, many times, and when I do, I listen to my team and tell my board about it, because that's what they are there for. The people around you in the prototype phase, when you are trying to make your ideas into a physical reality for the first time, are there precisely to push back and offer solutions that you might not have thought of. They are there to push you to pivot even when you may not want to. They are there to prevent another Theranos from happening.

The mistakes that Elizabeth made are common among young innovators. They are the mistakes of youth and ability. Elizabeth was a passionate and compelling storyteller—such a talented brand evangelist that she converted herself and everyone around her who was supposed to provide a reality check. She even turned her engineers into true believers, and engineers don't take anything on faith. And because of this, and yes, because of youthful pride, Elizabeth never had to pivot, and so her dreams of a world where people could access their health information directly and easily went up in smoke, along with $900 million of her investors' wealth.

Riding the Shark

Even though it may be difficult for many geniuses to admit, the prototype phase is designed to show you what doesn't work as much as it is designed to show you what does. Whenever you move from art to part, things will inevitably change. You will discover that this technology or that material or even your whole idea doesn't work. At that point, the goal is to determine why something doesn't work, because obviously, until you know what's not working, you can't fix it. You have to put a lot of thought into your failures during this process—it's not strong enough, it's too heavy, or it's too expensive—because that's what building out is about. Fixing what's wrong. Because of this fact, you can run into a million moral dilemmas you never saw coming.

Failure is an inevitable and critical part of the innovation process. You have to run up against walls, because getting over the walls makes your product better. You have to be able to admit when you're wrong, learn how to eat your hat, and then carry on to the next dilemma. The prototype phase forces you to dial in on what you're developing to make it workable. And sometimes you discover that what you originally envisioned just isn't workable no matter what you do. But maybe something else you discovered along the way is. So you pivot.

I call this riding the shark—the ability to jump into a problem and go where the facts take you. When you realize that your original idea won't work, you have to admit it, as Elizabeth Holmes should have done. Only then can you ride the shark to places and things that often wind up better than your original conception of things and certainly more purposeful, because it actually works.

Five Key Phases of a Prototype Plan That Will Really Happen

There will be lots of plans flying around as you approach this phase of the process, many of which will never get

> Failure is an inevitable and critical part of the innovation process. You have to run up against walls, because getting over the walls makes your product better.

done. If you really want to lay out a road map that you can realistically accomplish, try to think about the prototype development in terms of these five key phases.

Phase 1: Concept Phase

In the concept phase, you will explore an idea for a new or modified product or a new use for an existing product to determine if the idea or concept is worth pursuing from company, business, and market perspectives. Spending time and money on the concept phase saves time and money during later phases and can prevent future patent infringement lawsuits or the placement of a product on the market that is not medically reimbursable.

During the concept phase, you explore the product idea from multiple perspectives, including market trends, user acceptance, competition, product risk, regulatory path, patentability, and product reimbursement. You also determine whether the product idea makes sense from a strategic standpoint. You explore if the product idea already has an existing patent(s). If a patent exists, are there ways to design around the patent to come up with a new patent? If a patent does not exist, can the idea be patented and, if so, in which geographies?

Very early in the concept phase, you need to assess the market size and demand for your product idea. If there is not a strong need for the product, you should focus your efforts on coming up with a better idea that people actually need and want. Analysis of the competition is also important during the concept phase. Are you competing with large competitors who have a major share of the market? If so, you will need to develop a market penetration strategy. In my medical device space, I also need to understand if the product has an existing reimbursement code. If not, I need to come up with a reimbursement strategy if I want to pursue that product idea.

During the concept phase, you evaluate the potential risks to the users of the product. If the product is considered high risk, is the company willing to take on the risk of potential lawsuits from patient deaths and/or serious injuries? Finally, you assess how the product will be regulated in various geographies. Is a clinical study needed to commercialize this product idea? Clinical studies can add three to five years to the product development timeline and often cost millions of dollars.

Phase 2: Feasibility Phase

During the feasibility phase, you will either confirm or discredit the assumptions you've made about how your product will function and be used. You will be answering questions like the following: Will I be able to make the product reliably and have it function through multiple lots and multiple assemblers? Which prototype out of the multiples I have created works best? A lot of testing goes on in this phase until you have a version that is good enough for use and safe enough to go to market. When you achieve this, you have reached a point that is called "design freeze," when you stop adjusting and release your invention to be fully verified and validated.

Phase 3: Verification and Validation (V&V) Phase

During the verification and validation phase, you will test all of the mechanical and functional attributes of your product and bring them all in line with established specifications. You will perform simulated-use testing and also test how well your product holds up under extreme conditions. It's a lot like that commercial with the paint swatches in a field to test how it weathers. You have to know how the conditions your product will encounter will impact its use and appearance. You also need to test the biocompatibility of your device. It must meet international standards established by the US FDA as well as European and other global requirements established if you desire to commercialize worldwide.

Phase 4: Regulatory Phase

In this field you will research the regulatory standards pertaining to the classification of your device: I, II, or III. Then you will test as necessary and make critical adjustments to ensure that you meet the regulatory standards for your product as set by the FDA, CE Marking, and others. The Verification and Validation (V&V) phase is designed to test the products that meeting the regulatory requirements for the jurisdiction where you plan to release your product commercially. For instance, the FDA has certain requirements for class I and II products. If your product is cleared under a 510(k), then there are guidelines established by the FDA from which you can create a submission and file directly with the FDA.

If your device requires an IDE or is a class III or a premarket approval (PMA) product, then you are best served in most cases to schedule a presubmission meeting with the FDA to ask specific questions to streamline your submission for subsequent approval of your clinical trial prior to gaining approval to commercialize your product technology. The regulatory phase should be treated as a business strategy rather than just checking a box. The FDA should be treated as a partner. The FDA supports new indications and de novo submissions that bring new and safer diagnostics and therapeutic devices outside the US. CE Mark is a second approval path that you should or will seek. In some instances, it may be your first regulatory filing. Gaining CE Mark is transforming from the MDD (Medical Device Directive) to the MDR (Medical Device Regulation) as of the publishing of this book. Selecting a notified body (NB) to partner with your business is also a serious and critical step to make. NB's are your business partners in Europe and share business risk with you. Be selective in choosing this group. Select an NB that has experience in your domain and that may have worked with your competition. This will ensure that they know the risks and regulatory requirements for your domain.

Phase 5: Commercial Phase

Once you reach commercial phase, you no longer have a prototype. You now have a finished product that you can offer to the market. Initially, a limited market release is performed, followed eventually by a full commercial launch. More to come on this phase later.

The Care and Feeding of a Healthy Board

When you're building a board of directors, it's important to choose people who are experts in fields that are of particular concern to you and your business idea. You want people who understand your technology, your market, financing, your regulatory requirements, your business, and the law. You can draw on your counsel of experts to fill you in on areas where you may not be as proficient as someone who does it for a living. Think of your board as advisers who can guide you and contribute their special knowledge to the team.

It's also critical to have a board of directors that is willing and able to speak truth to power and push back. The worst thing you can do is put together a board with a lot of yes-men and women who aren't involved, aren't informed, and are just there to cheer you on and tell you what you want to hear—or even worse, let you tell them only what *they* want to hear. You run into trouble when you have a board that would rather just cover their eyes, bury their heads in the sand, and wait for a sale at all costs.

I have built several boards, and I'm currently on a few right now, and I can tell you that no matter what side of the table you're on, nobody likes surprises. You have to have a clear, open, and honest relationship with your board. Keep them informed as you are going through the development process. They are an integral source of support when things are going well and also when you're running into difficulties. If you keep your advisers apprised of events, they'll support you.

When you tell your board, "Hey, this isn't working out, but I think I'm on to something else," they will be more likely to go along with you, to follow your lead if they have been in the loop all along. They can also help you make the difficult decision that things just aren't working and it's time to make a major change, such as remove the present CEO and replace him or her. There have to be people around who will grab the cord on the bus and say, "That's it. Stop the bus!"

There's a saying I've heard and that I live by: "The board meeting happens before the board meeting." Seven to ten days prior to your board meeting, send out the board slides. Then call each board member and ask to review any questions they might have right there over the phone. Clear the air then and not at the board meeting. The board meeting is not for surprises. A good board meeting should be chaired and last two to four hours if managed properly before and during the meeting.

There will be special circumstances when this does not hold true. In 2015, when I was president of Blockade Medical, I was hosting one such board meeting. I had several board members as well as one observer attending. An observer is generally not an official designation. More commonly, it's just a seat at the board table reserved for someone other than the board member. As an observer, you are an active participant in board meetings but don't vote on any board matters and in some cases need to step out of meetings (typically to protect attorney/client privilege, which covers board members but not board observers). I had a person who was designated as an observer. It was the CEO of a publicly traded company. He would normally call in to the board meetings and listen in. On this occasion, he attended the meeting to observe in person.

We conducted the "open session" of the meeting and then took a break. He asked to sit in on the normally privileged "closed session," and we all obliged and continued with the closed session. During this part of the board meeting, the observer asked to speak and then proceeded to state that his company would, at that time, formally make an offer to acquire the company. You can imagine how disruptive this was. This statement resulted in a discussion of valuation, which we were not in a state to discuss. Our company was a technology play. We were not going to be valued on multiples of revenue or EBITDA. Our valuation would be derived by the revenue multiples and based on our intellectual property (IP).

In any event, we engaged in discussion over the next few days, resulting in a letter of intent (LOI), and initiated due diligence. Over the next ninety days, we had several meetings. The LOI did not result in an acquisition and left the company in need of more cash to continue with commercialization globally. As stated earlier, the board of directors do not like surprises, and for a very good reason. This surprise did not end with the result we might have hoped for.

If Things Feel Wrong, They Probably Are

Early in my career, I got a call from a friend who said he knew about a doctor and a business guy who were looking for an engineer to help them develop a hemorrhagic stroke device. This sounded like a great opportunity for me to join and lead a startup at an early stage, so I went to meet the team first in Boston and then down in Charleston.

I first met with the physician, who was very smart and kind of quirky but strictly legit. I recognized him; I'd seen his kind before plenty of times. He was the real deal. Then I met the CEO. He did not seem as transparent. I had never seen his type before. I didn't recognize him. He was this supersmooth, superexpensive-looking dude in Paul Stewart custom suits with one of those accents that could be from anywhere or nowhere. He was an enigma to me, and this made me nervous about going into business with him.

I didn't have a good feeling, but I was twenty-nine years old and wasn't really in tune with my instincts yet. I didn't trust myself.

I remember wondering, though, "Why is an alarm going off in my head whenever I'm around this guy?" But then I thought, "This is an opportunity to break away from working for a big company and be part of something major. If I have to make some sacrifices, so what? I'll survive."

So, I joined the company. And for the first six months, the business seemed legit—everything felt right. I was traveling a lot internationally with the CEO, introducing our idea to the key influencers in our field. We were going first class all the way, and that part felt really right—until it didn't. When we were in Buenos Aires, we stayed at the Alvear Palace, where the lobby, I swear, looks exactly like the Hall of Mirrors at Versailles. And I am not exaggerating. These places were the finest in the world. In Zurich we'd stay at the 175-year-old Baur au Lac, where Elton John and George Clooney stayed when they were in town. In Paris, we were at the Hôtel de Crillon, the French Downton Abbey overlooking the Place de la Concorde, where the suites cost thousands of dollars a night. This was in the late 1990s.

I was having the time of my life. I didn't realize that this level of accommodation was unusual. Well, maybe I did a little bit, but I didn't let myself think about it, because it wasn't my deal. I didn't make the hotel reservations. Our CEO picked his hotel accommodations by looking at a book called *The Leading Hotels of the World*, which I believe may have been the very first hotel rewards club.

Things started to get a little squishy when I got back from a trip from Charleston. Our controller came to me with the CEO's expenses on his credit card and asked me if I had been on the trip with him to Charleston. Our CEO didn't file expense reports. He just directed the controller to pay his credit card. My eyes almost popped out of my head. He had bought art, and furniture, and jewelry. He had spent a whole additional week in Charleston and in Kiawah Island after I had left.

I said nothing. I mean, who was I? These guys were, like, fifteen years older than me. I was the new kid on the block, and I wasn't involved with the financing of the company. I was an engineer. My job was head of technology. And our technology was an embolization coil system that was saving lives. I was only responsible for my own operating budget. I had nothing to do with the CEO's expense reports. "It's not your problem," I said to myself. "Don't get involved."

Another flag went up during our second-round financing. A company in Santa Rosa, California, made an offer to acquire us for $175 million. That deal would have made me a lot of money—a fortune for a thirty-year-old guy in 1999. We were all over the moon, dreaming of houses and boats and Ivy League. We took a trip to Santa Rosa to meet the leadership and present to the shareholders. And I thought everything had gone very well.

We were on the veranda drinking wine, reflecting on our success against the backdrop of a gorgeous sunset over the vineyards, when their head of R&D came up to me.

"Dave," he said, and he looked way more serious than the circumstances warranted. "I wonder if you would consider moving to Santa Rosa?"

"Well, sure, once we close this transaction, I'd be happy to consider relocating," I said sort of quizzically. I didn't quite understand why he was asking me this. Of course we would consider Santa Rosa after they acquired us. "Santa Rosa is awesome," I said, trying to lighten the moment. Maybe I had misunderstood his tone.

"Dave, I don't think you understand," he said then, looking even grimmer than before, which I wouldn't have thought possible. "We're not going to acquire you guys, and I'm going to tell you why, although I find it sort of hard to believe that you're not aware of this. We have been doing due diligence on your company and on your CEO, and we could never acquire a company like yours. Not with your financials and the type of person you have as a CEO. If I were you—and this is only a suggestion, but a firm one—I would resign from that company and go to work for somebody else, or join our group and develop a neurovascular platform for us. Things aren't going to go well for Micrus with things like they are now."

"This is shocking," I said a little louder than I should have, because I was watching all those millions of dollars along with everything they were going to buy me suddenly slipping through my fingers. "I've got a stake in this company."

"One hundred percent of something that's worth nothing is zero," he said. "You've got nothing. I would get out, Dave. Get out while you still have a reputation and before you wind up with less than nothing." He walked away, and my guts immediately started to twist. Alarms went off, red flags waved, warning flares exploded in the night sky. But I'm thinking, "You know, the technology works, so someone is definitely going to acquire it. So let's just get to market, and we can figure out what to do after that." I couldn't let go. I couldn't admit defeat. I had already put in so much time and effort, and we were so close. So, I ignored my instincts.

Thanksgiving turned into Christmas turned into New Year's, and our CEO finally realized the deal in Santa Rosa wasn't happening, so we went out again for another round of financing. We had a series of meetings over the next three months. Then, finally, in April of 1999, while we were at a meeting in New York City at the American Society of Neuroradiology, we got word from an investor group that they were really interested in investing in us. We were finally back on the finish line! Maybe I had been right to keep silent and let this deal close. And then … I got a call from my guy inside and the shoe dropped again.

"We're not doing the deal. You didn't pass due diligence," my friend said. "Now, I am going to say something to you, and if this ever gets back to me, I'll deny it, or I'll sue you for libel, or both. Your CEO runs that company like it's his own personal piggy bank. This is why we're not investing in you guys, and this is why no one else will either. If I were you, I would leave as soon as you can." This guy was telling me the exact same thing that the head of R&D from the company in Santa Rosa had told me on that magnificent veranda at sunset.

About a week later, I was back in California, and a large, thick envelope was delivered to my front door in San Francisco anonymously. There was no return address. I opened it up, and inside were the goods on our CEO—hirings, firings, divorce decrees, a collection of various social security numbers, a lease for a Range Rover in somebody else's name, a receipt for a $10,000 Patek Philippe Calatrava watch, and a diploma that wasn't from Harvard after all. My heart pounding, I packed it all back in the envelope and took it with me to work.

"Holy shit," said my friend Jack, the cofounder, who'd gone to engineering school at MIT and medical school at Tulane and was fond of superlatives. "What are you going to do with this information?"

"I don't know," I said, because I really had no idea. The whole thing was creeping me out.

"This is what you're going to do," Jack said, "and you're going to do it immediately." He was speaking bluntly and kind of scaring me, which I think was the point. "Get yourself some gloves before you handle those documents any further. Then, go to Kinko's and make fifty copies of every piece of paper in that envelope. Then go to a bus station with lockers, and put copies in a bunch of lockers—not just one, but several, get me? Then I want you to draw a map of where the lockers are, and give that map to me. That way, if you get murdered, people will know who did it because I'll have the evidence in the lockers, because I have the map."

"You're nuts," I said, but I did make fifty copies, and I put them all in my safe at home, and then I did nothing but worry for weeks. Should I drop the bomb? Should I listen to Jack and stash these in multiple lockers at the bus station? If I drop the bomb, how would I drop it, and on whom? I had no idea how to handle a situation like this, but I couldn't just put it out of my mind anymore. The time had come. I had to do something.

I finally worked up enough nerve to go and talk to the chairman of our board of directors. I told him what I knew and showed him the contents of the envelope. Predictably, he got really mad. I mean, who could blame him? I'd just told him the company was not worthless but very much at risk.

"Who told you this?" he asked me, sounding a lot like he was ready to blame the messenger. "Where the hell did you get these documents?"

"I don't know," I said, "but does it matter? Three people have told me this very same story, and now this arrives. I think we have to do something."

"Listen, just keep this under wraps for a while until I can get to the bottom of this. We are so close; let's just keep the train on the track, Dave. For God's sake, don't run us into a wall in the final mile!" That's when I realized that while the truth will set you free, it doesn't make you many friends.

Startups are like small towns. Everybody knows everybody else's business, and it's impossible to keep a secret. Somehow it got back to our CEO that I had been asking questions about him, and next thing I knew, he was calling me into his office.

"Have a seat," he said and started to pace. "Dave, today is your last day at the company," he said, finally sitting down in his own chair and facing me. "I'm terminating you."

"Really? What did I do?" I asked. I knew why he was firing me, but I wanted to hear the story he'd come up with to tell everybody else. I wanted to listen to him lie to me.

"You know what you did," he said and pulled out his flip phone and checked it, as if someone had called, to avoid my eyes. "You have been violating my integrity and my

> Start-ups are like small towns. Everybody knows everybody else's business, and it's impossible to keep a secret.

authority, and I can't have that around the office. This office needs to be unified, and you've just vilified me, so I'm terminating you."

And in that moment, a moment I had been dreading and fearing; a moment when I should have felt outraged and frightened, all I felt was relief. It was like this incredible weight that I hadn't even known was there had suddenly been lifted. All my stress was gone; so was that feeling in the pit of my stomach that had been there since I first took the job. I felt right with the world again, and that honestly was worth more than all of the millions I might have made if that company had been acquired. In that moment, I realized that my gut and my intuition were right and true. Everything my parents taught me growing up—right from wrong, good coaches from bad coaches, ethics, morals—it had all stuck. I was a guy who knew how to do the right thing. I could trust myself. I hadn't known that before. I knew it now. And it was very liberating.

By the way, my story did wind up having a happy ending eventually. The CEO finally did get fired after a lengthy lawsuit ending in an arbitration agreement settled at the American Association of Arbitrators in the famed TransAmerica building in San Francisco. The view from our negotiating room actually looked out at Alcatraz Island. Eventually, a CEO came on board who knew how to manage the company ethically. The company went public, and the stock I had retained in the company multiplied. So that is the final lesson I learned: justice usually prevails in the end, one way or another, but sometimes, you just have to be patient.

THE PROTOTYPE VIABILITY QUIZ

Here are five questions you should ask yourself before you go out with your prototype and make it an actual product for sale on the open market:

1. Manufacturability: Can you make it? Is it straightforward to assemble/manufacture? Are the materials readily available? Are they cost effective or too expensive?

2. Ease of use: Is it easy enough to use? Or does it need to be used by Dr. Golden-Hands? It's just a prototype, but can you envision the prototype in use? Is it intuitive to use? If it is not, then it may not be safe. If it's not safe, then you might injure

someone—causing another problem.

3. Functionality: How does it function? Does it perform as you expected it to? Is it too big, too stiff, too fragile? Is it the right size to be compatible with existing standard systems?

4. Marketability: Is there really a market for this, or is it not really solving a problem yet?

5. Profitability: What price are you going to charge? Will you be able to make a profit when you sell it? What is your anticipated gross margin?

THE STRANGE HISTORY OF THE PACEMAKER

Nine things out of ten don't work. The tenth one will pay for the other nine.

—Wilson Greatbatch

The pacemaker was accidentally invented by Wilson Greatbatch, an inventor with more than 150 patents to his name. Born in Buffalo, he attended Cornell, where he studied radio engineering and became a professor at the University of Buffalo. While working in his home lab in Buffalo, he was working on a heart rhythm recorder when he found a way to electrically stimulate a heartbeat.

A patent for the implantable pacemaker was granted in 1962, and in 1970 Greatbatch founded Wilson Greatbatch Ltd., which today makes lithium batteries for pacemakers. Greatbatch himself, however, never moved into his corporate headquarters, preferring instead to continue his research at home in his garage workshop.

EVERY SUCCESS IS BUILT ON FAILURE

There are no environments where you're only going
to win because life just isn't like that.

—Bobby Orr

W hen you're an entrepreneur and an innovator, failure is not really failure in the way that we generally think about the term. All innovation is a process of trial and error. Unless you make mistakes, you'll never arrive at the perfect solution. In this sense, failure is more like a teaching moment, an inextricable part of the process of experimentation that every new discovery must go through. In my company, failure is an integral part of success.

Since every new invention or improvement begins with a theory, that theory has to be tested. Is the material strong enough? Flexible enough? Does the solution actually function the way you anticipated? Every big idea depends on untested theories, and not all of them are going to work once you pull them off the drawing board. Failure is really a process of refinement. Through trial and error, theories become foolproof solutions, or an opportunity to realize that you have been barking up the wrong tree.

Close but No Cigar

At the beginning of my career, I was part of a team that was developing an interventional approach to treating brain aneurysms. Prior to interventional methodologies, you had to have open surgery on your brain, and this approach was obviously much more invasive than our idea. We knew that coils could work in other parts of the body, so why not in the brain? The market was new. Just one company had figured out a way to safely deliver and detach a coil in the brain. That was Target Therapeutics. They developed a product with a physician called the Guglielmi Detachable Coil, or GDC. It was brilliant. They had been acquired by Boston Scientific for $1.2 billion.

Target was successful in their own right before GDC had been developed and commercialized. But the GDC solidified their success and catapulted the interventional treatment of brain aneurysms. Many engineers and physicians saw the validation of the therapy and had been growing market potential. Until then, nobody had invented an effective and safe delivery method to deploy the coils in the brain without potentially causing an ischemic stroke.

We decided to develop our own new delivery system for an embolization coil that no one had tried yet. The market was flooded with intellectual property filings, but nothing like this was on the market. It was a minefield developing the product, though. We needed something tough but still flexible enough to go into the femoral artery in the groin and then be able to track across the heart and up into the brain, where it would be deployed into the brain aneurysm.

We decided to use a diode laser to activate a light-sensitive polymer that would soften in the heat being produced by the light absorption and release the implant into the aneurysm sac. The trouble was, the fiber-optic system, which we had housed in a tiny acrylic tube, turned out to be too fragile. It had worked great in our bench models, but we hadn't allowed for the tortuosity of the arteries in the neck and skull. The journey from the femoral to the brain was of a higher impact than we had anticipated, and when we went to deploy the device in a human, the tube was cracking before the coil could be deployed. We didn't know this at the time. This cracking caused light diffraction, which prevented the laser light to get to the polymer and soften it so it could release the implant.

This was not a small problem. This was the whole enchilada.

I discovered this critical design problem while a patient was literally on the table; the fiber-optic system failed, and the coil did not detach from the delivery system. The neurosurgeon, Dr. Lylyk, remained calm. He was like James Bond under pressure, this guy. Unflappable. Dr. Lylyk also understood this was a new and potentially superior technology than Target's GDC system, and he wanted to be a part of developing the device. When the coil failed to release, Dr. Lylyk looked up from his work on the patient, told me what had happened and calmly asked me what my plan B was. I gulped, and then told him to torque the device and break the implant off the pusher, which he did. The procedure was successful; the patient was fine. I was a wreck.

I reported back to my team, and we all went back to the drawing board. We built better bench models that more accurately reflected the torturous conditions that our device had just encountered inside the body. Then we began attempting to create a fiber-optic system that could endure those conditions without breaking. We had been counting on a light-sensitive polymer that could heat up to 150 degrees Fahrenheit, at which point it would soften and detach the implant. That system wasn't flexible enough to withstand the curvature of the human anatomy, making it impossible to get into the middle cerebral artery. The device cracked, and when it cracked, the light became diffracted and never went to the detachment zone.

Now we had to figure out a way to deliver the same degree of heat without using the diode. All we needed was heat, and it didn't matter how we got it there, so long as it reached a sufficient level to soften the polymer. What the team came up with was electrical energy. Two small wires were attached to a heater coil. This new system worked in the improved bench models, and the team was back on track but had lost valuable time. The technology had changed, too, so new patents had to be filed. Then we rushed to test out our new idea in the human body.

Of course, this big a pivot also meant that we had to inform the board of directors. We explained to the board that we had a system that would only work in lower locations in the brain but couldn't reach above the carotid syphon. We couldn't distribute a device that could only treat 50 percent of the aneurysms, so we had to change our delivery system to something more durable.

The board wasn't happy at all, which was understandable under the circumstances. I could have made them happy once more if I'd had the chance, but the CEO tried to use this as a way to fire me. He said to me that I had failed because I didn't meet the requirements of the task, and I told him I knew he'd been embezzling money behind everybody's back. Mic drop.

Then he fired me for questioning his integrity and violating his authority. Whatever. That was his MO. He always fired everybody when things didn't go his way. He didn't react well to failure, which is a fatal flaw when you are leading an innovative startup.

A Team That Has Failed Is Better Than a Team That Hasn't

Firing me was honestly the best thing that CEO could have done for me—and the worst thing he could have done for the company. Firing a team because they made a mistake is the hugest mistake of all.

Obviously sometimes you have to fire somebody for incompetence, but not after one or even two or three mistakes, because mistakes are a part of the success process. You can't fire people for trying new things and failing or you will never discover anything new at all.

When you hit a wall and you're trying to figure out what's next, you need the team that hit the wall with you to figure out how to get around it on the next pass. When you fire everybody and put a whole team of new folks on board to fix a mistake they didn't make, it's like starting from scratch. It costs you valuable time and knowledge. You lose the whole history of what you have learned from the trial-and-error process that is at the heart of productive innovation, and you may run the risk of making the very same mistake again and again.

> You can't fire people for trying new things and failing or you will never discover anything new at all.

It also puts your whole team on notice that failure is not tolerated, and without failure, no success is possible. When you have a risk averse culture like this, you won't have to worry about firing anyone; people just start to quit, and then hiring becomes the bigger issue.

You Have to Be Able to Adjust in Real Time

Just like a football game, you can plan strategy all week long, but when game day comes, things are going to unfold a little differently than you expected. Maybe the field is wet. Maybe the other team opts for a passing game instead of a ground game. Maybe your quarterback gets injured. You have to be able to respond in the moment if you want to win games. You rely on your plans, but you adjust as you go, and you don't consider that a failure.

The same is true with innovation. In my field, when you actually go into the human body with a new device, sometimes stuff happens that you couldn't have precisely tested for. Scientists and engineers expect that this will happen. This is why you have first-in-man studies. It's the only way to understand what actually works efficiently and what doesn't once you're doing things for real. When something doesn't work, you adjust. You have to be able to switch it up. Be light on your feet. Be willing to let go of the plan and try something else. Don't misunderstand what I'm stating here. Safety, biocompatibility, sterility are all proven. How the device reliably performs and the durability of the device takes a few attempts, though.

Pivoting midstream is hard for engineers and even harder for companies. When you don't hit your design requirements, when you exchange technology, this can cause problems in your valuation. You've built a development plan that gets delayed. This, too, impacts your valuation. You may have signed a technology agreement with an outside company for stuff you aren't going to use anymore. The board gets upset because they just want stuff to work. It's hard to admit that you've made a big mistake and have to radically switch up your game plan, but you have to do it. You have to be transparent, and you have to be able to pivot, or you'll be out of business.

How to Hire People Who Know How to Fail and Pivot

Assembling a team that is agile and confident enough to try new things in an R&D lab begins for me right at recruitment. I like to hire people who have experience with failure and understand it as a part of the process of winning. I like hiring former athletes for this reason, because they know what it means to train and prepare and then lose and get up and play another day. They understand that there isn't a player on earth who goes undefeated. They know that if you are going to win, you have to play, and if you play, some of the time you're going to lose.

I like to hire people who are competitive, because they will push the envelope in order to win and they are less concerned that sometimes when you push the margins, you go over the edge. Hiring people who have a familiarity with losing in the real world results in an agile and courageous team that is able to expect failure, make mistakes, and rebound readily from them with new and better ideas.

I also like to hire and work with former military or people who have been in law enforcement. Why? Because these folks are disciplined and fundamentally team driven. They have faced fluid circumstances with high stakes and have had to adjust in real time, often in life-or-death situations, as a unit. In my industry, it's the doctors who are really taking most of the risk, but they will expect me and my team to be there and be prepared to help them adjust if a problem occurs. You and the doctor both have to be able to think while a patient is on the table, listen to what the doctor is telling you, and be able to respond appropriately to help the doctor overcome the problem.

I also hire a cross-functional team who come from a variety of specialties, cultures, and backgrounds, because I like a diversity of opinions. I like to have people contributing all different points of view, because that gives us more bites of the apple. I will have six points of view about how to get around a problem. If you have a team that has all the same experience and background and cultural orientation, you get five answers that are all the same.

You Can't Teach Good Character

Innovation as a general rule is a little like the Wild West, where it's easy to make up the rules as you go along. That's why I want people on my team who already have a sound moral compass—people who I know I can count on to do the right thing. You can't teach good character; it has to be there already.

That being said, I like to hire young people, because you have an opportunity to guide and shape their point of view. There's less resistance, no bad habits. I also like to work with people who are nice. This sounds simple, but it's a big deal. Not everybody is nice, especially under pressure. Successful teams are supportive teams that come together to help out the weakest link in the chain, even when the stakes are very high. I want people who are empathetic, patient, and generous with their time. I want people who are fun to hang around with.

I also look for people who have volunteered in their communities or coached a kids' team because these activities require patience. Inventing new things requires patience by the bucketful. These are all qualities that in my field make for an agile team that is not afraid to fail and knows how to win even in the face of adversity.

> Successful teams are supportive teams that come together to help out the weakest link in the chain, even when the stakes are very high.

Learn How to Communicate Failure

When you make a wrong turn or something doesn't quite work out the way that you planned and you're forced to pivot, you have to be able to talk effectively about that with all parties concerned. You have to be able to talk to your team, your investors, your board of directors, your CEO and explain to them why something didn't work, why you made the decisions that you did, and what you intend to do about it. As a leader, you have to be able to discuss failure in a productive way, and so does your team.

If you don't handle failure competently and fearlessly, you are going to have a team that is afraid to fail and afraid to tell you about it when they do. And that is something every entrepreneur should fear, because if your team isn't honest with you, you aren't going to have the first idea about what's going on in your business. And the buck stops with you.

If you don't have periodic design reviews and communicate the results to the team, then the team doesn't know whether they have failed or succeeded. You have to include the team in all relevant conversations. It's very important to keep everybody fully informed about what's at stake and what progress is being made. Budget constraints are an important piece of information as well. This is why a lot of startups fail, because leadership doesn't know how to share the unvarnished truth and instead keep their teams laboring in silos, in the dark about the project as a whole.

If your team members can't talk to you and to each other about where they went wrong, if people try to hide things when they don't go right and leaders try to hide failure, then growth is stalled, because you can't learn anything from the experience and trust is completely eroded. And as I have said many times in my life and in this book, where there is no trust, there is no team.

Addressing failure with transparency, admitting that you were wrong, is never easy. It takes a special person to be able to communicate that they just spent millions of other people's dollars on a wrong turn. It takes an even braver soul to convince them to give you $6 million more to fix it. But that's in the job description when you're an entrepreneur. You and your team have to be able to make mistakes, and you have to be able to tell people about it, even people who aren't going to like it very much. Otherwise you wind up like Elizabeth Holmes and Theranos.

When you're an innovator, you're in essence selling a beautiful vision that exists mostly in the future and largely in your mind. So, you have to talk a lot about success. People like winners. They want to hear about the amazing things you're going to accomplish with their money, how you're going to change the world together, and how much money you are going to make them in return. Think about the scale of Elizabeth Holmes's pitch. She promised to make her investors billions of dollars by democratizing medicine and eradicating premature death from the face of the earth. That's a pretty huge claim. Still, people bought it.

Investors love dreamer entrepreneurs who think big. But those dreams have to be real. A time-travel machine is a huge and amazing idea, but nobody is going to invest in such a thing because it's make-believe. Dreamer entrepreneurs also have to be pragmatists who can get the job done. Which means that entrepreneurs have to tread a thin line between optimism and pragmatism at all times. This is achieved only when you temper your vision with truth and success with failure.

The Hockey Stick Effect Only Happens in Hockey

The hockey stick effect refers to a graph that shows a sharp rise of data points after a prolonged flat period. In a startup pitch, the hockey stick effect comes into play when you show a revenue curve in a five-year plan where year one is flat, maybe year two as well, and then around year three you commercialize and the curve shoots up suddenly and dramatically. Revenue goes from zero to, like, $20 million in a year. Investors love the hockey stick effect, for obvious reasons. It's like hitting a grand slam. And like a grand slam, it happens only once in a blue moon.

I had a conversation several years ago with a venture capital (VC) investor. This is a typical dilemma for VCs. After consideration of my pitch, he decided not to invest in a company I had cofounded for which we were seeking Series A funding. He didn't like the deal for two reasons. First, he was unable to put enough money to work in my company. I was seeking a $3 million round after raising $1 million at seed. He asked how much I expected we'd need to get to an exit. I replied, "One to two more rounds. All in, maybe fifteen million dollars." He told me that wasn't enough for him. This is not an answer I am used to hearing, so I asked him to explain. He told me that he knew he couldn't take 100 percent of each round due to his fund bylaws. He wanted to put at least $15 million of his fund to work on each deal. I didn't satisfy this requirement.

Second, he didn't see my deal at exit being big enough to satisfy his multiple. He said he swings for the fences on each deal. I thanked him for his honesty, but I was still a bit mystified. My deal wasn't big enough? The problem I was solving was certainly big enough to me. I told the guy that I understood his dilemma and shot back that out of every ten deals he did, he'd get one at 3x and one at 10+x. He told me that was right. Then I said he'd get his money back on three investments and crash and burn on five of his investments. "Those odds are terrible," I said. He said he knew but that he had a system that depended on his 10+ return multiple, and he'd been keeping his LPs very happy as a result. How do you argue with a guy who has happy shareholders?

Still, this really rubbed me the wrong way. I couldn't imagine putting any investors' money to work with odds like his, but it was apparently working for him. We parted ways, and we did get our company funded and were acquired at 5x for the first rounders and at 3x for the last money in. We still satisfied his second concern but never got close to his first concern. We only raised $12.5 million in total. My reluctant investor had been correct. We weren't a hockey stick. Even though he would have made better numbers in the long run, it just wasn't as exciting as watching those pucks fly from across the ice in year three.

Here's the problem with overly optimistic hockey stick projections. It can make even a success look like a failure, because you're over-promising in a situation where you will almost always under-deliver. Say you promise a sudden leap in year two and you don't hit your revenue in year two. Even though things may be going extremely well, you are now in a position to have to defend yourself.

What generally happens even in a successful venture is that growth is gradual. Even if you do realize the hockey stick effect, it's not a long handle. Maybe revenue goes from zero to $1 million or $2 million in year three, or even year six, and then grows organically from there. Double-digit growth year over year is phenomenal. These are more realistic scenarios, but again, investors like hockey sticks, so people tend to walk a thin line between optimism and caution in their presentations.

CHAPTER 5

HOW THE MONEY MOVES: FUNDRAISING ESSENTIALS FOR FLEDGLING ENTREPRENEURS

Want to change the world? Upset the status quo? This takes more than run-of-the-mill relationships. You need to make people dream the same dream that you do.

—**Guy Kawasaki**

Whenever you have a big idea that you want to bring to market, there are going to be resources involved that you may or may not be able to provide for yourself. Usually, even if you're just inventing a new Frisbee, you are going to have to raise additional money at some point to provide for your product's development.

There are lots of ways to raise money when you have a good idea, but there is an art and a science to how, who, when, and why when you begin to access funds and take on investors. In this chapter I'll try to give you an overview and a typical timeline for how the money moves.

Seed Investing

Seed investment is the initial funding used to begin creating a business or a new product. Obtaining seed capital is the first of the funding stages required for a startup to become an established business. Seed capital can be a relatively modest sum of money and might come from the founder's personal assets, friends, or family. It generally covers only the first essentials, such as a business plan and initial operating expenses. The goal at this point is primarily to obtain more financing, and that means attracting the interest of venture capitalists or banks. Neither is inclined to invest large amounts of money in a new idea that exists only on paper unless it comes from a successful serial entrepreneur with whom they are at least nominally familiar.

Founders' Capital Contribution

Self-investment is always the first step in financing your idea, getting it developed, and finally getting your new product to the market-place. The first line of funding should always, in my view, come from you and your partners. Founders must put in first money, which will be used primarily to file any IP, build initial prototypes, and prove that your solution works. It's highly possible, if things go well, that the founders continue to fund the development and growth of the company and never seek outside capital. This means that when you strike gold, you still own the whole operation.

The founders' capital contribution is important because it's going to make a difference to future investors that you have skin in the game. How can you expect to ask someone else to invest money in your idea unless you've invested in it first? People need to know that you believe in your concept enough to invest your own money into it.

If you build up a track record of success early on with your own funds, then you have a better story to tell. "I spent my own money on this because I thought it was so good, and guess what? My money is already paying off. I am on the right track, and I'm growing, and that's why I'm coming to you. I'm offering you an opportunity to get in on this great deal on the ground floor."

When you have first money in, there's less risk that you're going to go out too early and disappoint investors with early trials that you should have solved with your own money. When you have solidified your concept and your foundational pillars on your own money, you have a much better chance of being credible when you are ready to go out for investment dollars. You've come up with an idea, you bet your own money on the solution, and you are succeeding and growing. That's a very different proposition

> People need to know that you believe in your concept enough to invest your own money into it.

than "I've got an idea and maybe it works, but I need your money to find out if I'm right."

Family and Friends Funding

The first round of funding that generally occurs after founder investment is a process called "friends and family" or in some cases crowdfunding. Crowdfunding as an option has only existed since 2010, when the investment laws changed. Prior to that, the Securities Act of 1933, which was enacted after the stock market crash of 1929, prevented this kind of investment from happening. The Securities Act required that as an investor, you had to be able to prove that you could bear the risk of losing your entire investment.

In other words, you had to demonstrate for the government that you could afford your investment by having a net worth greater than the amount of your risk. You had to document your worth in three documents, get them all notarized, and then file them with the Securities and Exchange Commission. And then you had to acknowledge that unless there was fraud, you couldn't sue the CEO if your investment went south, because you had understood that risk when you made the investment.

The JOBS Act made it possible for small businesses to leverage capital from the general public without all of that pesky regulatory stuff. Now crowdfunding is possible, either through the many crowdfunding platforms that have sprung up since 2010 or through private investors. In my case, I crowdfund among my friends and family. There's a big advantage to friends and family funding. For one thing, it's much faster, because you've already been vetted. These are your friends and family, after all. They know you, and they already trust you. Due diligence has for the most part already been done.

Sometimes when you're in diligence with a VC or with another investor, they don't really know you that well, so you have to be vetted thoroughly. Even if they were referred to you, they are still going to want to check you out. This means drilling down a lot deeper than just checking out your LinkedIn profile, where people can say anything they want. And the friend that referred them doesn't have to be right. So investors who don't know you personally will take a thorough look at your credentials, and that can take a lot of time.

Ostensibly, your friends and family already know a lot about you. They've seen your triumphs and your defeats. They know where you're headed and how far you are along the way toward getting there. They see the cars you drive, the house you live in. They know what you are capable of. They know what you've achieved thus far.

Before I ever asked any of my friends or family for a dime, I already had a great résumé with some successes and a few failures, so my friends and family knew the difference between the two. They knew what it was like when I was on a failed venture, and they'd seen what it was like when I was headed for success. So when they saw how gung ho and passionate I was about the opportunity I was offering them this time, they trusted that it was a good bet.

Funding Blockade

The first time I went to my mom and dad for an investment was when I cofounded a medical device company called Blockade Medical. This was a business designed to develop coils that had already been FDA approved. I was an equal founder with colleagues from the industry. Our capital contributions initially were only, like, say $100,000 combined, but that was really just to get the company founded, file some initial IP, hire our first employees, and cover several months of business expenses. We initiated our formation documents, and then we had to raise the first $1 million to get the prototypes designed and hire a bigger team.

We went back to our own pockets to initiate this new round. We three partners wrote more checks, but we were still short of the $1 million needed to complete our operating plan and get to CE Mark. We went out and started pitching. We had a great deck. We were telling a very simple story. We were proposing a coil company that was off-patent and had a straight path to CE Mark and FDA approval. We intended to use our profits from our approved coil to go and develop other unique technology that we already knew the physicians wanted. All we were doing was making our good thing even better. It was an easy investment to make. I couldn't lose.

I went to my mom and dad and some friends of mine within the industry. My friend Marc had some friends in Europe we contacted. We reached out to physicians we knew well. So, our first big round was all self-investing and friends and family. And we hit our $1 million milestone easily. We avoided the implications of taking on big outside investors, which calls for an art and science that you need to learn and master.

Of course, it was a lot of pressure and very high stakes taking money from my friends and my mom and dad. I definitely didn't want to lose my folks' retirement! But I knew I had something that was going to work. I had few if any doubts about the return on their investment. And it worked out just like I thought it would, because it really was a foolproof model.

Use this story as a litmus test when you're considering inviting your family and friends to invest in your idea. Be as sure as you can possibly be in an uncertain world that what you've got is viable. Until then, self-invest so you don't lose anybody's resources but your own. In fact, make double sure, because as I have learned, sometimes things don't work out just like you planned. Sometimes, when you're riding high and thinking you've got it locked, a sudden wave crashes onto shore.

Chasing Neurvana

I had the experience of losing my friends' money on a company called Neurvana, which was a spin-off of Blockade Medical. When Blockade was acquired, the company that acquired it offered us the ability to spin out three products they didn't see as a fit for their business. I was allowed to act as chairman but had to hire the CEO and management team after the merger and acquisition (M&A) was concluded. Turns out the acquiring company thought that they had no use for these products. I thought, "Hey, that sounds pretty good. How bad could this be? I can't be the CEO, but I'll be the chairman, so I can provide vision and strategy."

I was hesitant about the guy selected as CEO because he was a commercial guy and a salesman to the core. He spoke in hyperbole, and it made me nervous. But I figured, "I'm current in the market, I'll talk to him, I'll be his partner, and we'll fix any problems we run into together." Everything felt solid, so I invited some friends to invest and also brought on a new investor whom I'd not worked with in this capacity before. We were moving fast, but it felt like a sure bet. What could go wrong? I couldn't see the downside but had some initial reservations about the CEO and new investor. However, it was blue sky all around me wherever I looked.

Eighteen months later, Neurvana tanked.

There was a myriad of reasons why. The new investor that came in saw this as a bigger opportunity than it really was, probably because he was an inexperienced investor in this sector and had been listening to the sunny forecasts of my CEO. Employees began to depart the company. At one point, I was removed from the board, and the remaining board leadership never communicated to the investors that I had been removed. Later, the company filed a lawsuit against me.

The first time any of my good friends who had invested in me heard about my removal was when the lawsuit filing went viral on the internet. This is when I started getting calls from my friends demanding to know why I hadn't told them about this. I mean, of course I couldn't tell them—I had signed an NDA and nondisparagement agreement when I was removed from the board and had remained as a consultant temporarily. And now I was being sued.

My friends said that the only reason they had invested in the first place was because I was chairman, and I should have let them know. And I understood their position, even though my hands had been tied. I ended up making about fifteen proactive phone calls to try to stem the bleed-out. It was not a good time. And to this day I still have about ten friends who as yet haven't talked to me, which makes me very sad.

What did I learn from this? Well, first of all, don't hire the wrong people, even if it's convenient to do so. If your gut says no, don't do the hire. I had a lot of responsibility in owning the transition for Neurvana, and I met with some opposition among the new leadership of both entities. I think the CEO saw this as an opportunity and he leveraged it. He was an opportunist, and I knew that, but it just seemed so easy letting him be the CEO rather than going out and hiring somebody from the outside. And honestly, I was worried about hurting his feelings, which I know now was stupid. The easiest thing to do, the path of least resistance, is rarely the right thing.

I also learned not to take the wrong money. We had an investor who was much more ambitious than I was about Neurvana's scale, and he was new to creating an investment fund. This was his first rodeo. He saw this is as a much larger opportunity than it really was, and so he made the wrong decisions. This proved fatal.

Take money that shares your vision, because the money will have to stand behind you and support your decisions. Don't reach out to friends or family until you've got the right people already involved, because the wrong people can and usually will tank the whole thing. In the end the lawsuit was dismissed. The judge saw the claims made by Neurvana to be meritless. It cost me some reputation, but worst of all, it cost investors money.

Crowdfunding Platforms

There are a lot of new companies reaching out to new investors on crowdfunding platforms like Indiegogo, Gofundme, Fundable, Kickstarter, Rockethub—there's a million of them now. Very respectable projects have evolved from investors gathered in this way, and I think it's a very good thing for startups. And the numbers you can raise are not small. Pebble Watch raised over $10 million on Kickstarter. The open gaming console Ouya raised over $8 million on the same platform. Bitvore raised $4.5 million on Fundable, so it's a serious game and also a great way to test out your product with your target consumers.

Equity Crowdfunding

Equity crowdfunding is a new phenomenon. It wasn't legal until President Obama passed the JOBS Act in 2012. The JOBS Act was created with entrepreneurs and startups in mind, hoping to allow small businesses to advertise opportunities to their target markets. Known as "general solicitation," public equity crowdfunding campaigns are a great way to build out your network and invite grassroots money to support your mission. It's an easy way to invite friends and family to invest, to build excitement, and to communicate with your base. It also makes for an incredible and extremely millennial brand story that expresses your company's core values. Crowdfunding in medical device development has not become a stable or relied-upon method to financing your idea.

Angel Investors

The term *angel investor* actually comes from the Broadway theatre, where affluent "angels" would swoop in at the last minute and provide the resources necessary to keep productions running that would otherwise have closed. In business, angels usually come in the beginning stages of a startup, when the risk of failing is high and when more traditional investors wouldn't be prepared to invest yet.

Angel investors are usually entrepreneurs themselves and often give for reasons that go beyond pure profit. They may have belief in a particular cause, want to mentor the next generation of entrepre-

133

neurs, or want to accelerate development of new technologies like, for example, cancer treatments or green energy. In addition to funds, angel investors are also valuable sources for mentorship, guidance, and contacts, as many have had success in your sector and will make the connections you need to help your venture and their investment gain a return.

Today, many angels are starting to organize themselves into clubs, generally ninety-nine members or less, who all support the same mission. In Irvine there is a group like this called The Cove. These clubs usually have about ten billionaires, then there are forty guys who are worth more than $500 million. There's another forty worth less than $500 million, and they meet as often as needed to see pitches face to face. There are no public listings of angel investors or angel clubs. You can meet angel investors in several ways but mostly through referrals from a trusted contact. There are also angel investor conferences, where startups in a particular area of concern are invited to pitch in person to the angels.

Series A (and More) Financing—Typically Your Venture Capital Round

This is the round that provides for the resources you will need to verify that your design meets all the design requirements and you get to design freeze. This round should get you to a fully working and functional version of your product—not commercial stage but a functional version. This means that when you get to this round, you are starting to talk about some big money proportional to your earlier rounds, because you're going to have to start paying for bigger-ticket stuff.

In order to get to design freeze, you've going to have to hire a team. This means salaries and benefits and reduced equity. Then you have to provide an office space for these people to work in, so you'll have to lease a building and establish a footprint for your business. In my field, I have to build a small area that has a controlled environment called a "clean room." The scale of the expense really depends on your needs. How big a team will you need? What kind of an environment will you require to produce the product? How complicated is your solution, and how long will it take you to get to design freeze?

In many cases this is where the V&V of the product occurs, or the concept of the product may need more feasibility testing leading to a design freeze and V&V. This round is funded in most cases to get regulatory approval of your product. Later rounds can be what is necessary to get to market: to commercialize and build next generations of products.

You go to venture capital firms, often called "strategics," private equity firms, and once again high-net-worth individuals or angels to find Series A rounds.

How to Build a Compelling Deck

A deck is a sales tool that is used to organize and guide a twenty- to twenty-five-minute presentation in front of your potential investors. Most pitches will be longer. However, having a deck that you know you can get through in less than thirty minutes is a plus. Any good deck should be able to express all of the relevant information in under a half hour. There's inevitable Q&A during the pitch, but you should be able to get though the first two slides uninterrupted. In most cases, this deck is used to get you a second meeting and not to close the deal on the spot. That rarely happens, if ever. This deck could also be used for a "fast pitch," which is a common kind of pitch competition, so you need to be mindful of time always. Fast pitches could range in time from three to fifteen minutes.

I have been writing my own decks for decades, and they usually work pretty well. They work well because I've learned through personal experience and continuing to seek out from others what is required or requested from potential investors. So I thought I would offer my outline for how I lay out my slides and what is important about each to get you started on your own.

Slide 1: Vision and Big Idea—Three Minutes

This slide presents the grand vision and the crux of your solution. What's the big idea and your "why"? Think of this as the first three minutes of a movie. You have to explode onto the screen and offer a hook that will keep your audience in their seats for the rest of the movie. This slide should create excitement or even fear and gain and keep your audience's attention. A friend of mine, Oren Klaff, has taught me quite a bit about pitching. He's written two books on the topic: *Pitch Anything* and *Flip the Script*. I refer to both still. I recall an interview he had that made an impression on me. According to Oren, we all have a "crocodile brain" that evaluates anything new based on three questions: "Do I kill it, do I eat it, or do I have sex with it?" If you can't get your pitch past this investment decision gatekeeper, then good luck closing the deal.

We've all been in that pitch meeting where you see eyes glaze over as people start to think about much more important things like sports, Facebook, and food are at that moment. You might be giving them the most beautifully logical reasons why they need you, but without the sex appeal, you're dead in the water.

Slide 2: The Problem—Three Minutes

If you aren't solving some problem in the world, you are going to have a long uphill climb with your business. Use this slide to talk about the problem you are solving, describe who has the problem, and humanize the impact that it's having on their lives. You can talk about the current solutions in the market, but don't spend too much time on the competitive landscape on this slide—you'll have a chance to do that on a later slide. Ideally, try to tell a relatable story while you are defining the problem. The more you can make the problem real and the pain points palpable, the more your investors will understand your business and your goals. Do not make this a slide about how much money you're going to make your investors. You must be solving a real problem, not just making money.

Slide 3: Target Market and Opportunity—Three Minutes

Use this slide to expand upon who your ideal customer is and how many of them you think there are. In other words, describe the total size of your projected market and how you position your company within that market. If you can find the data, investors will want to know how much people or businesses currently spend in the market to get a sense of the total market size. This is where you tell the story about the scope and scale of the problem you are solving. If it makes sense for your business, you'll want to divide your market into segments that you will address with different types of marketing and perhaps different product offerings.

Be careful with this slide, though. It's tempting to try to define your market to be as large as possible. Instead, investors will want to see that you have a very specific and reachable market. The more specific you are, the more realistic your pitch will be and the more your potential investors will trust the information you are giving them.

Slide 4: Your Solution—Three Minutes

Finally, you get to dive into describing your product or service with this slide. Now is your opportunity to explain how customers will use your product and how it addresses the problems that you outlined on slide 2. You'll be tempted to move this slide closer to the beginning of your pitch deck, but try to resist the temptation. This is classic storytelling where you build up the problem and describe how bad it is for lots of people. Now your product or service is coming to the rescue to help solve that problem.

Most entrepreneurs are very focused on their product when instead they should be focused on their customers and the problems those customers face. Try to keep your pitch deck focused with this format, and you'll tell a better story. If possible, use pictures and stories when you describe your solution. Showing is nearly always better than telling. You want to minimize your potential investor reading the words on your slide as you're pitching, so make your copy brief and your presentation vivid.

Slide 5: Revenue/Business Model—Three Minutes

Now that you've described your product or service, it's time to talk about why this whole brilliant idea of yours is going to make people money. You will answer questions like "What do you charge?" and "Who pays the bills?" For some businesses (content sites, for example), advertisers pay the bills instead of users, so it's important to flesh out the details here and help your potential investors understand where the money is coming from and how your model works.

You can also reference the competitive landscape here and discuss how your pricing fits into the larger market. Are you a premium, high-price offering or a budget offering that undercuts existing solutions on the market?

Slide 6: Traction Road Map—Two Minutes

If you already have sales or early adopters using your product, talk about that here. This deck can also be used for later rounds of investment. Investors want to see that you have proven some aspect of your business model. This really helps investors understand that you have reduced the risk for them. Any proof you have that validates that your solution works to solve the problem you have identified is extremely powerful.

You can also use this slide to talk about your milestones. What major goals have you achieved so far, and what are the major next steps you plan on taking? A product or company road map that outlines key milestones is also extremely helpful here.

Slide 7: Go-to-Market Strategy—Two Minutes

This is where you illustrate how you are planning to get your customers' attention and what your sales process will look like. Finding and winning customers can sometimes be the biggest challenge for a startup, so it's important to show that you have a solid grasp of how you will reach your target market and what sales channels you plan on using. Use this slide to outline your marketing and sales plan. You'll want to detail the key tactics that you intend to use to get your product in front of prospective customers. If your marketing and sales process is different from that of your competitors, it's important to highlight that here.

Slide 8: Financials—Two Minutes

Investors will expect to see your financials. These generally include a sales forecast, a profit and loss statement, and a cash flow forecast for at least three years hence. In the med-tech business that I'm in, everyone wants to see your five-year operating plan, kick the tires, and check out your hockey stick projections. That's a must and a good exercise. Keeps your finger on the pulse of your business.

The first two years of revenue projection is the most accurate. Years three to five are a total pipe dream, but still, investors love to see the hockey stick uptick at year three showing you'll dominate the world or market segment. That's always confused me. I'm sure it has a psychological effect on the investor that shows you're ambitious.

For your pitch deck, you should never include in-depth spreadsheets that will be difficult to read and consume in a presentation format. This really slows down your pacing, and you risk losing your audience. Limit yourself to charts that show sales, total customers, total expenses, and profits. Revenue, EBITDA, and cash flow are key metrics most novice or sophisticated investors will want to see.

You should be prepared to discuss the underlying assumptions that you've made to arrive at your sales goals and what your key expense drivers are. Remember to try to be realistic. Investors see hockey stick projections all the time and will mentally be cutting your projections in half. If you can explain your growth based on the traction you already have or support your conclusions with data from a similar company in a related industry, that is extremely useful.

Slide 9: The Competition—Two Minutes

Every business has competition in one form or another. Even if you are opening up an entirely new market, your potential customers are using alternative solutions to solve their problems today. On this slide, describe how you fit into the competitive landscape and how you're different from the competitors and alternatives that are on the market today. What key advantages do you have over the competition, or is there some "secret sauce" that you have and others don't?

The key here is explaining how you are different from the other players in the market and why customers will choose you instead of one of the other players. Differentiation is very important. Differentiation can be as simple as that your product is just easier to use. Ease of use in medical devices can translate into a safer procedure, reduced procedure time, and better outcomes.

Slide 10: Investment and Use of Funds—Two Minutes

Finally, it's time to actually ask for the money. That's why you're doing this pitch deck, right? I know—I said that this pitch deck isn't about actually getting funded in that moment, but you do want a second meeting. Your five-year operating plan will detail the money spent, milestones you expect to achieve, revenue you expect to achieve, and, if all goes well, a path to profitability or cash flow positive. You will break down that operating plan into segments separated by key milestones, which could be product design freeze, a regulatory approval, or a first-in-man use of your product.

You will decipher which one of those will create value to the business. You will then inform your potential investor of the milestone you've selected, and be sure you add a buffer and explain that this is "how much" investment you're seeking in order to achieve milestone A. The operating plan will be detailed to show the use of funds to achieve the value-creating milestone. Details such as headcount to be hired, maybe capital equipment to be purchased, testing to be completed, suppliers to be used, and so on, are also important. Be as detailed as you can be, because potential investors do need to know how much money you are looking for and what you intend to spend that money on, as explained in slide 8.

Most importantly, you need to be able to explain why you need the amount of money you are asking for and how you plan on using the money. Investors will want to know how their money is being used and how it is going to help you achieve the goals you are setting out for your business. If you already have some investors on board, now is when you should be talking about those other investors and why they chose to invest.

Slide 11: The Team—One Minute

With this slide, you'll talk about who is on your team and why they are the right people to build and grow this company. What experience do you have that others don't? Highlight the key team members, their successes at other companies, and the key expertise that they bring to the table. Even if you don't have a complete team yet, identify the key positions that you still need to fill and why those positions are critical to company growth. It's important to state that this team has to be complementary and not just a couple of people that are smart and excited about the solution you have envisioned.

You can't have two CEOs or two CTOs. People with similar skills can have different titles and roles, but they need to be getting

143

the work done. Sometimes the founders are in the leadership roles and sometimes they aren't. I don't mind seeing this, as it shows me that the founders know their place in the process and are not just taking a C-level title out of entitlement. I've been part of companies where the cofounders had lesser titles but were high contributors. I've been part of the reverse as well. Investors are trusting that you've sorted this out before you pitch them. They invest in people first.

Slide 12: Exit Strategy and Partnerships

I shy away from this discussion on an initial presentation. I would not present this slide until you know that you have interest from investors.

Mezzanine Financing

This is later or last round to keep the company going or pivot the technology before an exit, such as an M&A or IPO. So-called mezzanine financing is sometimes necessary to support a business into its introductory phase. This is usually available only to businesses with a track record, and even then only at a high rate of interest.

Washout Financing

You are only in the market for washout financing when things have gone terribly and horribly wrong, because you will be lucky to see fifty cents on the dollar, which means your investors will have lost half their investment. Also, there could be management changes during this phase of financing. The reason milestones were not achieved and the valuation isn't where expectations were is because of the management. They couldn't figure it out.

This is how a washout can become necessary. Say I've raised $1 million in my seed with a valuation of $2 million—postmoney valuation is $3 million. I did a series A; I raised $3 million on a premoney of $6 million because I hit my milestones—post of $9 million. Then I did a series B; I raised $10 million on a premoney of $18 million (because I hit my milestones)—postmoney of $28 million. After my B round, I have a postmoney valuation of, say, $28 million because I've been hitting my milestones. All is looking good.

Then I do a clinical study, and I miss my clinical and technical end points. The FDA says, "No, you need to do this, this, and this." So now I've got to raise another $25 million, and then I'm fine. So I go out and start to raise an additional $25 million so I can complete the operating plan, but before I can do that, somebody gets to market with my technology first. I still have IP and a safe product, but it's not proven yet to show it works clinically. What should my valuation be? I missed milestones!

Now nobody's as excited. Am I toast? I go to a VC and I say, "Hey, listen—my technology works well, but now I have a competitor." The VC will say to me, "Okay, but what's the valuation? You didn't complete the milestones in your last round, and now you have competition. You are no longer first-to-market either. I will give you the $25 million, but I'm debating whether this is a flat round or if we should lower your valuation." You are no longer in control of your valuation. This also means that all of the people who invested $14 million with a postmoney of $28 million now will see further ownership dilution at a minimum or have the value of the stock they purchased end up worth less per share.

You will take a deal like this only if you still want to bring your product to market and have the potential of even having those previous investors break even or get, like, fifty cents on the dollar. You would only do washout financing in dire circumstances, when something has gone gravely awry and you are trying to save at least some of your investors' money.

CHAPTER 6

ASSEMBLE A WINNING TEAM

I hire people brighter than me and then I get out of their way.

—Lee Iacocca

What would Steve Jobs have done without the Woz, or Guy Kawasaki? What would have become of Warren Buffett without Benjamin Graham or Charlie Munger, or Wilbur without Orville? It's hard to imagine, because innovation doesn't happen in isolation. No one person can do the whole job of bringing a new thing into the world. Even with the greatest invention since sliced bread in your hands, without the right team, you will probably flame out. In my experience, it isn't until the right team comes together at the right time to work together with commitment to find a solution that you get to *Eureka!*

While you have begun to reach out and make relationships to get yourself this far, once you have created a viable prototype, the focus really has to switch from the product to the people. Who can be your CEO, who can create and express a narrative that will motivate people to come on board your train? Who will raise the money? Who will lay out a sales plan and execute it? Who will troubleshoot regulations? Who will build and refine the product? Now, before you can go any further, you need to assemble a winning team. Are you the CEO? The commercial guy? The architect of the technology? All of the above? You can't do all of these things well. You can do some, of course. You'll need to surround yourself with the people who complement you and shore up the gaps.

My senior staff in 2018 had thirteen people on it, and for all intents and purposes, I'm the coach. I've got running backs and defensive ends and tackles and linemen on my team, and everybody has to do their job, because just like any sport, innovation is a game that is won or lost on the strength of the team, not its individual stars.

Teamwork Builds Bonds That Can Go the Distance

I stuttered from the time I was four years old all the way through college, and in an odd sort of way, it was my stuttering that taught me a lot about being a team player. Here's a good example of why I say that. So, I grew up in Leominster, Massachusetts. My dad was a tool and die maker, and a lot of polymer and plastics molding companies were there—DuPont, Union Carbide, Nypro. A lot of injection molding. Very industrial.

We lived in South Leominster, near Lancaster. I'd ride my bike all the way to Lancaster and back to a variety store to buy baseball cards. And on the road to Lancaster, just before the place where I'd stop to buy my baseball cards, was where James Bentley lived. Every day, twice a day, I would pass the Bentley house, and I would see James hanging out front, playing by himself. I knew him a little. We'd played at school together, but we weren't really friends. James didn't have any friends. He was kind of a dirty kid, and the rumor was that he had bugs on account of the fact that he had gotten a crew cut one winter. After that, all the kids just figured they'd had to shave his head to get rid of the bugs. We all have been guilty of unreasonable snap judgments, and kids can be the worst, because they have no filter.

So James used to get picked on, and while I wasn't ever one of the guys who picked on him, I didn't make much of an effort to be friendly either. I mean, I judged the book by its cover. I went along with the crowd. I was like, "This kid's a freak. He's dirty; he doesn't have any friends. There must be something wrong with him. He must have bugs." That's what I thought, until fate intervened, and I was the one getting picked on.

There was this Puerto Rican kid, taller than me, who made fun of the way I talked because I stuttered. He talked with a Spanish accent. He thought *I* talked funny? One day, while we were playing in the schoolyard during recess, that kid picked on me one too many times, and all I remember is punching him in the face and knocking him to the ground like Frankie beating up Flick in *A Christmas Story.*

Long story short, I went to the principal's office, he went to the nurse's office, and I got a week's detention, and somehow, on the very same day, so did James "Bugs" Bentley. There I was in detention, making clowns out of construction paper with James, and because I had been forced to spend some time with him and get to know him a little bit, I realized he wasn't such a bad guy after all. James came from a challenging background. His folks weren't around a lot, so he didn't look as put together as the rest of us. But he didn't have bugs. The rumors hadn't been true. James was a pretty funny and smart guy, and he made way better construction clowns than I did.

In detention that year, I learned an important lesson very early. I learned that when you get to know your teammates, when you actually spend time together, understand a little something about where they've come from, what they've gone through, you realize that they are generally much nicer people than you thought they were. It's harder to snap judge somebody you actually know. This is a lesson I carry with me to this day into my innovation teams. I try to encourage my teams to spend time together, to share where they come from, because mutual understanding is the glue that holds a winning team together.

The next year, my parents put my sister and me into private Catholic school across town, where they imagined there were fewer bullies, and I didn't see James Bentley anymore. The kids in my new school were very smart, and everybody was pretty solidly upper-middle class, so it was a culture shift for me. And I stuttered, which was, you know, a stumbling block. My parents had thought they were helping me out by putting me in a more affluent district, where there were no more tall Puerto Rican kids to ridicule me, but I had already carved out a pretty good niche for myself in the old neighborhood thanks to baseball and ice hockey. Now I had to do it all over again on a brand new team.

And once again, because I was a good athlete, I found my place in the new school. Because I was on a team, people were able to understand that I wasn't just some weird kid from the other side of town who stuttered but instead, a pretty funny guy, and a good athlete, who helped their teams win.

As a kid who stuttered, being part of a team helped me communicate in ways other than verbally, which was a challenge for me. I communicated by executing and by doing my job on the ice, on the baseball diamond, and in the classroom. I was never an individual, per se. I was always a team player. The same holds true in my businesses and on all of my development teams today. We win as a team. We make mistakes, have setbacks, and learn as a team every day.

Dave's Team Rules

I had a coach in high school who in some ways taught me more about business and about life than he did about baseball, although he taught me a lot about that too. He summed up his approach for all of us in a set of team rules that he had posted all over the locker room. I've adopted my own version of those team rules, which I feel are at the heart of how big ideas really happen. I use them as my true north, and I expect my teams to live by them. I have them posted in nearly every war room on every project I undertake, to remind everyone of my rules of play.

DAVE'S TEAM RULES

1. Go all in, and no matter what happens, no quitting.

2. Everyone has a talent and a specialty—respect it.

3. Be humble in victory, confident in defeat, and quiet about both.

4. Straddle the line between the learning zone and the out-of-control zone.

5. Get to know your teammates as people, not just as positions.

6. Stop talking and listen.

Go all in, and no matter what happens, no quitting

My very first rule of play is that you have to be all in. A winning team requires that each and every member of that team is as invested as you are in winning. I demand passion and belief. I want people who really believe the problem we're solving is a problem, because those are the people who will really strive to find not just one but multiple solutions.

For me, it's really a lot like sports. Major league baseball players, for example, aren't there just to win the game; they are there to win the World Series. From opening day forward, they are focused on the end goal with passion and commitment. That's the same level of commitment I expect from my teams. When you are dealing with real-world problems, you need real-world solutions, and the only way to come up with those is to have a team of people who are all in.

Everyone has a talent and a specialty—respect it

Nobody is good at everything. Tom Brady couldn't be a defensive lineman. He's a quarterback who can read the defense and get rid of the ball superfast. So what do you want from Tom Brady on your team? You want him reading the defense and getting rid of the ball superfast. You don't ask Tom Brady to be a kicker or a lineman; you put other people around him to do that stuff, and you free up Tom so that all he has to do is that thing he does so well. It's the same principle on my innovation teams. As a team leader, as a coach, I recognize what people are great at, and then I structure the organization so that they are free to do just that.

Just like on a football team, my innovation teams will always include people with different but complementary skill sets. I have to have people who can design catheters, people who can design coils and stents. I have people who know how to write design plans, people who are great at testing protocols, and people who are highly skilled at handling regulatory procedures and speaking with notified bodies and the FDA, as well as people who could sell sand on a beach. It's your job as a leader to figure out who does what best and then galvanize that talent into a well-calibrated machine that can win consistently in all situations.

Here's an example. I have a woman working for me who is the best neurovascular catheter designer in the whole world. Everybody knows she's the best. Michelle is a total rock star. She is not, however, the world's best program manager or the world's best communicator, according to her peers. She can drive our quality and regulatory people crazy because she's usually late and doesn't provide updates so there are inevitably surprises. She doesn't stress about project planning. She doesn't stress about giving employees reviews. She used to actually upset doctors during demonstrations because she would disagree with them and tell them what she thought they needed. But Michelle is a great listener. And because she can listen, she is able to take in the critical information, understand the problem immediately, and then go design the perfect catheter for the customer. When the product arrives, it is a beautiful thing.

One of my directors was, like, "You've got to do something about Michelle—move her to a different position or get rid of her, because she's irritating the doctors." I told him, "Why would I let the best neurovascular catheter designer in the world go? I would have to have my head examined. A move like that wouldn't be good for anybody, and definitely wouldn't benefit our results." Still, he had a point. The things she wasn't good at still needed to be done. So, just like when my coach made me a catcher so we could

> It's your job as a leader to figure out who does what best and then galvanize that talent into a well-calibrated machine that can win consistently in all situations.

benefit from our talented pitcher and win the state championship, I built a team around my movie star to pick up the slack on things that aren't in her wheelhouse and freed Michelle up to spend her time just being an amazing and talented designer.

Every team must contain a variety of people with myriad skill sets who focus on their own talents and who understand and respect the talents of the other team members.

Be humble in victory, confident in failure, and quiet about both

One of my heroes, Wayne Gretzky, said, "Say nothing when you lose, and say less when you win." This really sums up my approach to team success and team failure, because both are just a moment in time. At the end of the day, whether you win or you lose, you still have to move forward.

Of course, when you fail, you have to talk about it with your team. You have to discuss what went wrong, figure out why you failed, and devise new systems and strategies so it doesn't happen again. You have to remind your team that failure is part of the success process and that it's not going to stop you. It's just a teaching moment that the whole team can learn from together. But this should be a private moment. It's not for public consumption, which would only make it harder to pick yourselves up and move on.

Be democratic in success. Share the glory; spread it around throughout the entire team so it can raise all boats. Talk about your success together, and figure out what you did right so you can solidify systems and repeat the win. By all means celebrate. Take your victory lap, and then move on.

Success, too, should be private. Don't overemphasize your victory in public, because at some point you are going to lose again. Nobody is ten for ten. Not even Tom Brady. There's always going to be someone else, some other person, some sleeper who sneaks up on you because you were celebrating and took your eye off the ball, who figures out a better, more efficient way to get it done.

Straddle the line between the learning zone and the out-of-control zone

I want a team of engaged, worldly individuals who aren't afraid to walk the line and walk the line between the learning zone and the out-of-control zone. I think you can't emphasize enough the value of being worldly. You don't have to be worldly to get where I'm at, but it certainly hasn't held me back, and it's given me a lot more confidence. The places I've been, the things I've seen because of my job are experiences I wouldn't trade for anything. I've been to the US Open, the French Open, the Tour de France, the World Series, the Masters, the Super Bowl, Champion League Soccer, Boca Juniors, and so much more. If you can think of an iconic sports event anywhere in the world, I've been there, or it's on my list. It's an amazing part of my life, and it's brought sports back into my life in a way that I never expected. And it is all because I am an entrepreneur and part of the global community of innovators all engaged in trying to come up with solutions to make life better, just like me.

These experiences have really added to my ability to connect with great people all over the world. They have really expanded my horizons and my possibilities for success, both personally and professionally. This is why I want people on my teams who aren't satisfied with the status quo and who want to keep growing. I want people in the out-of-control zone, because they expand possibilities and help the whole team grow.

 ## The Learning Zone–Going Where Angels Fear to Tread

In 2019, I was in Turkey, twenty miles from the Syrian border. ISIS was there and only two short years ago had been cutting off people's heads and broadcasting it on YouTube. Yet, I felt totally safe. I felt safe because I have been in danger zones throughout the world many times: Israel, Argentina, Brazil, Colombia, China. This kind of an extreme situation is actually in my comfort zone, as it has to be in my field. I have enough experience to know when I'm safe and when I'm actually in danger. I feel safe also because of the good people I'm with when I am traveling in scary places, at dangerous times. There are good and great people all over, but you have to be able to recognize them when you see them and then listen to their advice and do what they tell you to do when you are in their country. That is how you stay safe in a war zone: having friends you can trust and the good sense to listen to them.

My engineer, on the other hand, was way out of his comfort zone. He was thinking, "Man, why are we doing this? This is dangerous; why are we here?" He thought he saw ISIS around every corner. I told him, "Well, we're here to help a patient. We're here to train physicians. And I trust that the physicians who invited us wouldn't put us in harm's way." After this trip, my engineer will have put this kind of travel in his learning zone. And after traveling with me for a while longer, he too will be just as comfortable as I am out on the wire.

I tell my teams to constantly look for ways to expand their minds. Take that seminar, watch that documentary, travel to Turkey ten miles outside of Syria. Go to Shanghai where no one speaks English and figure out how to get a cup of coffee. You're an expert on extrusion? Okay, great—now go take a seminar on how to make nitinol or something else you don't know how to do yet. Do something that is out of your reach, because once you start to tackle it, it comes inside your learning zone and eventually becomes part of your comfort zone. What was once impossible becomes possible, and even easy.

Get to know your teammates as people, not just as positions

Everybody's different, and you've got to acknowledge those differences if you want to be a good team player. What you do, who you are, where you've come from, how you grew up, and what your life experiences are all contribute to how you solve problems. Innovation is about problem solving, so it's important to get to know your teammates as people in order to understand what they are contributing professionally and how best to work with them.

As I mentioned previously, I stuttered growing up. Knowing this will teach you a lot about how I approach problems. My production manager grew up in Vietnam and immigrated to the US when he was a teenager. My quality and regulatory strategist was born in Taiwan and came to the US as an infant. His parents didn't speak English. An engineer I recently worked with was born in Japan and didn't come to America until he was sixteen. All of us have very different ways of approaching a problem based on these facts that have impacted our perspective. When you're on a team, you have to know these things about your teammates because they build understanding. Understanding builds trust, and if you don't have trust, you don't have a team.

I'm an entrepreneur. Entrepreneurs have to build strong teams. Whether you're in a growth mode or you're in a public company, it can get pretty strategic in the Shakespearean sense of the word.

There are swords behind every rampart. Everyone is gunning to be number one; everybody has a secret agenda. You have to be able to trust your team, and trust begins with education, communication, and understanding.

Stop talking and listen

I learned this all-important lesson from a friend of mine back in 2002. I was in a weird place. I'd been fired maybe a year earlier from an early-stage company and my ego was bruised. I had come down to Orange County to work for a competitor. This company was looking for people like me, so it was a good opportunity to see what the business was all about and meet some folks. The team was full of stars, which is always problematic. After having been there for about a year, I was at a project update where an engineer was presenting a potential solution to a problem he was having.

I didn't get off to a great start. Early in the project update, I stopped listening and started talking. I stopped the speaker cold. I was like, "You know, you don't even have to go any further. I don't need to listen to any more of this. I already know this isn't going to work. I did this at my last company, blah, blah, blah." I thought it was my moment to shine. I certainly showed that I was leaps and bounds ahead of the presenter. People listened, but no one said anything. My friend, who years later became the CEO of that company, pulled me aside afterward.

"Dave, why did you have to shoot your mouth off like that?" he asked, clearly exasperated with me. "Why did you have to be such a know-it-all? That engineer hates you now. Everybody in that room hates you."

"Really?" I said, kind of stunned by the criticism. I certainly hadn't meant to make people hate me. "I thought I was helping. Come on, he doesn't hate me."

"Oh yeah, he hates you, all right," my friend said with emphasis, the way you do when you know you're right on the money. "He hates you, and he wants to kill you now. Why couldn't you just shut up and listen?"

And you know, that was the moment when I saw my own reflection in the eyes of the people in that room. My friend had shown me something extremely important—something that only a friend can share. He had shown me myself through the eyes of other people. And he was absolutely right. I was acting like a jerk, I wasn't listening, I was shooting my mouth off and alienating people who could have taught me something, and now I had made enemies when I had wanted to make friends.

That lesson changed my life and my career, and so I try to teach everyone who plays on my team about the importance of listening and paying attention. I ask my team members to remember that they don't know everything, that they can always learn something if they give people their undivided attention. If you insist on thinking that you know everything, you will wind up a jackass of all trades and a master of none.

Building a Launch Team

When we formed MindFrame in 2007, we had the foundation for great product development and innovation at Intersect Partners. The incubator CEO had formed a team comprising technical people who could provide financing, design, product development, manufacturing, and intellectual property filings as well as create a regulatory and quality strategy and get to market through a commercial distribution strategy. These resources were a combination of consultants and full-time employees from the incubator effectively on loan to MindFrame through a contract services monthly fee. This move made it easier in some ways and harder in others.

All I had to do at that point was complement that already stellar team with people who could assemble the prototypes, test the prototypes, get feedback from customers with a team of clinical experts, and develop a clinical strategy. We held off on hiring our own crew of full-time people until such time as we gained our first regulatory approval, which we knew would be in Europe upon CE approval. Sensible solution, right?

We had made this decision to outsource a team for financial reasons. It seemed safer than to incur staff expenses month to month at an early stage of the game. But as many bargains do, our cheap solution turned out to be efficient in the short term but more expensive in later phases of the company's operating plan. The initial low-risk monthly fee in the concept and feasibility phases was reasonable but became a very high price to pay for security and convenience after regulatory approval. Later on, as we entered the commercial phase, we should have hired a dedicated team of top talent and had capital to spare that would have increased our runway to market.

When we realized this, the partners sat down and identified all of the key skills we needed to enable the company to cut the cord from the incubator and become independent. Based on that assessment, here is a list of the core staff we needed to hire, which can be useful for anyone who is trying to build out a skeleton startup team to take them to launch.

KEY PERSONNEL

1. **R&D Engineering Manager:** This position designs the product and adjusts it to meet customer needs.

2. **R&D Technician:** This position assembles and tests the devices designed by the R&D engineering manager.

3. **Quality Assurance and Regulatory Affairs Manager:** This position ensures a robust quality system with documentation and design control. This person should also have an understanding of what indications are needed for regulatory approvals and developing testing protocols to get the product into regulatory compliance.

4. **Clinical Manager in Europe (or International):** I am in the medical devices space, so we needed someone who could determine which centers

to test the product at first in Europe and then later in other countries. This person also has to know how to comply with the regulations of the hospitals, enabling that first-in-man study.

5. **Manufacturing Manager in the US:** This position develops the processes needed to make a device reliably and eventually scale production to meet the demand created by the commercial team.

6. **Commercial Manager in Europe (or International):** This person creates the go-to-market strategy upon CE Mark approval (or other approvals) of our device abroad.

7. **General Manager:** This is a position like the team coach, and I often function in this capacity. Whenever I worked as the general manager, I wound up wearing a lot of hats. I was an R&D engineer testing devices and presenting to the physicians to get feedback to improve or confirm our design. I was a clinical manager meeting the key physician influencers and selecting centers for FIM studies. I was the strategist on intellectual property ensuring we were filing patents on the novel features and benefits of the device and procedure. I was the evangelist professing to the neurovascular market what was next and why we needed this technology in our field to improve health care. I was a pitchman raising funds with our chairman and CEO to continue funding the

> business to meet our objectives. So obviously this is a crucial position.
>
> We hired all of these key people with efficiency and pieced them together like a puzzle. We were able to do this because we had autonomy. We were young opportunists with ambition, and we could call our own shots. We were all on the same page and incentivized with equity to succeed. This is exactly what you want in a launch team—people on the same page who will jump in and do whatever they have to do to get the job done.

Recruiting Kindred Spirits

Rather than conducting a traditional interview with people and asking them standard questions about their top five challenges and their top five successes, I try to get to know people personally.

I'll try to draw them out and have a conversation about what their life and background is all about. I do this by talking about my own background. I'll tell them, for example, that I just got back from a vacation with my kids, that they swam with the dolphins, and that it was an amazing experience. Then I'll ask them if they've ever done something fun like that. Or I'll tell them I went with my high school buddies to see the New England Patriots play a road game, and ask if they had ever done anything like that.

Then I listen, and I see if they can express their own history and experiences, how well they can reflect who they are behind the title, and what they think is fun. I am looking to see how well they can communicate, how well they connect with me on a human level, and how well they will fit in with the rest of the team.

Why is communication so crucial? Because if they do come up with a solution, they are going to have to explain why that solution works to lots of different kinds of people and connect.

They are going to have to explain how and why their solution works and field debate from people who will think that they're wrong. They will have to debate with people who will want to convince the room that their idea is terrible, and they are going to have to come back and win that argument. This is why communication is so crucial to me.

This is why I can't have people working just with their calculators and slide rules. I need engineers who can communicate. And you can tell by the way people communicate what their values are. Do they have humility? Do they have a sense of humor about themselves? Are they adventurous? Are they kind? Are they confident? Do they travel? Do they play sports? You can tell all of these things about a candidate just by drawing them out about their lives and listening.

Another crucial component for me is hiring people who are team players. I will ask people if they played team sports in school. I'll ask if they like working in teams. Athletics is a terrific training ground for the kind of attitude and values I'm looking for on my team. If you played sports in school, you already know about managing your own time, because if you play sports in high school or college, you have to know how to get your academic work done and still find time to train. They know about winning and losing because they have already won and lost as a team. They know how to be coached, how to work as a unit. These are all qualities that fit hand in glove with what I expect from the people who work for me.

My approach to recruiting kindred spirits involves drawing people out, getting them to talk about themselves, and really listening to what they are telling me about who they are, what they value, and how they feel about that. This is the way we get to know people in our personal lives, and work life is really no different. Get to know the person behind the résumé, and follow your instincts.

The Care and Feeding of Engineers: A Field Guide for Entrepreneurs

Most inventions that go commercial will inevitably be grappling with the problem of how to communicate with that rare species that nobody understands—the engineer. As someone who started as an engineer and now works as a CEO, I understand both sides of the aisle. I am therefore an expert in the care and feeding of engineers and am providing this field guide for entrepreneurs who don't understand engineers but need them anyway.

1. **Challenge them.** Engineers like to be challenged and love figuring things out. Let them.

2. **Train them.** Send your engineers to seminars and conferences, and invest in further schooling. Keep them learning, circulating, developing new skills, and honing existing ones.

3. **Liberate them.** Let your engineers be good at what they're good at, and don't force them into a corner or box. If an engineer is gifted at making catheters but challenged at finishing the project, then build a team around that engineer. Everyone is great at something. Build a team with great skills and talents. Don't force everybody to be great at everything, you'll wind up with a team of people who are only great a fraction of the time.

4. **Time limit them.** Give your engineers a goal with a deadline. Good is good enough. It doesn't need to be perfect. The sales team needs something to sell. Get a product to market that meets expectations and the marketing specs. Don't let your engineers tinker into eternity, striving for perfection. Settle for safe and good.

5. **Regulate them.** Give your engineers rules, and then let them play within the rules. In other words, don't move the goal line on them. Give your engineers a problem to solve and the specifications within which to solve it. Then leave them alone.

6. **Budget them.** Give your engineers a budget to solve the problem. Otherwise they'll invent something that maybe you can't ever sell.

The Silent Wisdom of the Domestiques

I like to bring speakers in to talk to my teams, and because I come from a sports background, I will often bring in athletes to talk to my teams about teamwork from their point of view. One of the most compelling speeches was from a guy named Frankie Andreu, who was a champion cyclist on Lance Armstrong's team when he won the Tour de France. Frankie has an interesting career as a professional cyclist because he wasn't the star, he was the worker, and as such, I felt he had a lot to teach me and my team about what it means to be a team player.

I learned from Frankie that a bike race involves a whole lot more than one guy on a bike. What most people don't realize is that bicycle marathons like the Tour de France are not solo sports. Those are actually teams that you watch compete. For every CEO cyclist like Lance Armstrong—the guy who gets to actually cross the finish line—there are seven vice presidents, called "domestiques," who are just as instrumental to the win but never get the glory.

Domestiques are a French invention. The Tour de France is a French race, so obviously they're the experts, and they figured out that a 2,200-mile race requires a group effort. And they were right.

Nobody wins the Tour de France solo. *Domestiques* is a French word meaning "helpers" or "workers," and they do everything from carry water along the course to cooking dinner to helping devise strategy to pacing the pack to taking on a competitor and wearing him down. Sometimes they attack and spread; sometimes they lie back and slow down the race.

Frankie says that when you're attempting to win the Tour de France, everybody has to be on the same page, and there has to be a strategy that everybody buys into. There are no individuals in the race. Lance Armstrong could never have won without those seven other cyclists helping him get the job done.

Before each day of the race, Frankie and his team would discuss strategy and plan their attack. Maybe today is a rest day and they just follow the pack. When you're on a bike for twenty-one days, you can't attack and spend that kind of energy every day. Maybe the next day will be an attack day, when you swarm a competitor or surround an opposing team and slow them down. When you're riding three hundred miles a day, you have to schedule out what you are going to do, how you are going to do it, who is responsible for what, and what the goal of each day is, just like in a business.

What's so thrilling about watching the Tour de France once you've seen it through Frankie's eyes is that you realize what you're seeing is teamwork that is so well synchronized that it can function silently, seamlessly, and at very high speeds. What's also so inspiring are the domestiques themselves—athletes who are so dedicated to their sport that they don't even care that they're not the ones who get to ride across the finish line as long as someone on their team takes the trophy.

CHAPTER 7

PLANNING AND EXECUTING A LIMITED MARKET RELEASE

Truth is like poetry, and most people hate fucking poetry.

—**Anonymous**

When I'm getting ready to launch a new product, before I go wide, I always do a limited market release (LMR). This means that I supply a controlled group of influencers that represent the range of my people in my target market. I'll go to them, demonstrate our solution, let them respond to me, and report back about their experiences.

A lot of inventors don't bother with an LMR. It can be costly, it slows you down, and you might not like what you hear. But I always do a limited market release because there are some very good reasons for doing so. An LMR gives you the opportunity to further refine your design, to make it perfect and unique while simultaneously building buzz about your new product.

An LMR also gives you the opportunity to educate the public about your invention or product and explain to people why they need your invention in their lives. Believe it or not, sometimes people don't realize right away why your invention is the greatest thing since beer and pretzels.

Test and Verify

Before you can bring a product into a limited market release, you have to test your product for efficacy and safety and carefully document your results, because you'll need them down the line. The demands on testing vary according to industry. In my highly regulated medical devices industry, for example, you need a big testing budget and meticulous strategy, because the FDA sometimes requires rigorous trials before approving a new product or therapeutic device. But even something like a new Slinky or Frisbee will need a reliable test and result phase because toys wind up in the hands of children.

The testing phase is crucial to successful development because you learn a lot about what doesn't work. You learn what could work better; you address the details that often get overlooked, like manufacturing costs per unit, shipping and packaging considerations, and overall appeal.

The testing phase is also a time when the entrepreneurs and the engineers on the team often clash, because their cycle and goals are different. The engineers' innovation cycle ends with an answer to the question, "How?" The entrepreneurs' cycle begins with it. Engineers and entrepreneurs battle over the product's design freeze all the time. Is it ready yet? Is it good enough? In my field, if it functions and is safe enough, then freeze it and get into a limited market release. Revise later. That's not how most engineers or commercial folks feel about the situation. They will develop until the cows come home, because that's what they do. They refine and perfect.

> The testing phase is crucial to successful development because you learn a lot about what doesn't work.

Confronting the Skeptics

In my business, an LMR always includes presenting at conferences, either by physicians on the podium or in front of doctors in a private technology suite, who can be a very skeptical and vocal audience. While I was with MindFrame in 2008, we presented our stent retriever to doctors around the world at the EuroStroke Conference before putting ourselves out on the open market. I can remember attending a conference in Nice, France, and I was telling a group of neurologists and neurointerventionalists that we have this new way to remove blood clots from brain arteries when patients have a large-vessel occlusion. I told them that it's going to be much easier and more reliable than what they are presently doing. I explained that we're going to put a stent-like structure into the occluded artery, trap the clot, and pull it out.

"You're out of your damn mind," a doctor at the meeting said abruptly. "I would never do what you're suggesting. That's going to cause so much damage. It's going to dissect the artery and kill the patient on the table."

Obviously, this startled me and everybody else who was listening. This doctor did not mince words. But I was ready for him. I was prepared, because I had been in an LMR phase for a while, and I had a device that I had seen work with my own eyes.

"No, doctor, that won't happen," I was able to say with confidence. "I've already heard of a similar procedure performed in the United States that will be published in a white paper soon," I said. "They partially deployed a stent 'off-label,' restored blood flow upon deployment, and then slowly removed the system as it was deployed— and it worked. The clot came out." I looked at him, waiting for his stern look to soften, but it didn't.

"I don't believe you," the neurosurgeon said. And he was German, so it sounded very official and dismissive. "But then again," he generously allowed, "this isn't even a problem for me. I rarely see patients who have strokes with large-vessel occlusions."

"And the reason you don't see many patients with large-vessel occlusion strokes is because you don't have a solution. We have one now," I said. Well, it was true. Unfortunately, my self-confidence only fed the dark energy in the room, and it grew.

"I do see a lot of stroke patients," said another doctor in the far back of the room. "I also don't see why this would work. Why would this help as opposed to hurting the patient more? I don't see it. No, I don't see it at all."

"Well, because we have a solution, and while I'm actually developing a better one right now, the current solution is working today in our clinical study. Doctors are already saving lives." I was trying to stress the life-and-death potential of this solution. We were talking about patients like my Pepe, and still these doctors didn't want to hear it.

"I don't believe you," the doc said, predictably. I mean, at that point, what can you say? They weren't going to believe anything that I had to say. I had to show them. And this is precisely why you do a limited market release. An LMR makes it possible to demonstrate for the skeptics out there—and they will always be out there—that there is a problem worth solving and that you have solved it successfully and reliably.

Before You Prove the Solution, You Have to Prove the Problem

When you have a game-changing technology, you're going to face a lot of disbelief out there from people who don't even think there's a problem. It isn't until you actually demonstrate the problem that people understand. Until then, you hear things like, "I don't believe you," even when the facts are staring them right in the face. The woman who invented windshield wipers couldn't sell the patent because nobody thought you needed a wiper on a windshield. It was just too far out of the scope of their understanding that one day we would be in sealed mobile environments and wouldn't want to open up the windows to put our arm out to clear away the snow in a storm.

Two months later, we went to a similar meeting, met with the doctors, and showed them the results of a pilot study that included more than fifty patients who had their blood flow restored by having had clots removed with our device. That's when they finally got the aha moment I had been waiting for, and we began to work with some of the neurointerventionalists present to perfect the device.

THE STRANGE HISTORY OF ROCK AND ROLL

Can't no man play like me. I play better than a man.

—Sister Rosetta Tharpe

It wasn't Elvis, or the Beatles, or even Chuck Berry, but Sister Rosetta Tharpe, born in 1915 in Cotton Plant, Arkansas, who invented rock and roll. Coming out of the gospel world, Rosetta was the first woman to stand out in front of her band wielding an electric guitar and rocking it out for commercial audiences. In her 1944 hit "Down by the Riverside," she literally shreds her guitar like a modern-day rock star.

While we don't often think of a form of music as an invention, it was in fact the very same process of combining elements that already existed in a novel way to create a new form that changed society.

Of course, Sister Rosetta's new invention didn't achieve mass adoption until the day that Elvis Presley saw Sister Rosetta, was attracted to her unique style of picking, then got up in front of audiences and played like Rosetta did. Back in those days, sadly it still took a white man to demonstrate the power of a black woman's invention, but fortunately for Chuck Berry, and Little Richard, and the Rolling Stones, Rosetta's innovation took hold.

LMR Tips for Marketing Novices

If you are new to the whole concept of marketing and would frankly rather chew glass than engage with your customer base, these tips are for you. They will help you to at least understand the goals and process of an LMR so you know if the somebody else you appointed to do your dirty work is doing it right.

Meet Incredulity with Credibility

Have you ever heard the statement, "There's no such thing as a bad idea"? It's not true. There are bad ideas. In fact, there are a lot of them. Remember that story about the guy who wanted to feed mayonnaise to the tuna to make tuna fish? Bad idea. This hopefully fictional guy should have first performed a voice of customer (VOC) to verify what the customer wants. He would have been stopped in his tracks. But after a VOC, you dial in the design specs, which should lead you to an LMR. The only way to present a new idea credibly is to take off your blinders and make sure the idea you've ultimately developed is a good one and works the way the customer wants it to. An LMR is an excellent opportunity to field harsh feedback, face questions, take a long hard look in the mirror, and know whether your idea is a good one or not. Once you know, incredulity is a paper tiger.

Use Your VOC to Plan Your LMR

This is when the VOC reviews that you exercised at the beginning of your development process will really come in handy. The marketing and design specs that you gathered then should provide the foundation for your limited market release plan. If all has gone well during development, the product will hit the goals you set based on those very early customer inputs. If the LMR verifies that you met the customers' expectations, then simply put, you nailed it!

How to Use Your LMR like a Free Marketing Campaign

A limited market release is by its very nature exclusive. It's limited. Not everyone can participate. It's the innovation equivalent of a velvet rope. You generate excitement by excluding people. When the buzz is generated, people who haven't been invited suddenly want to know why they were excluded and how they can participate. A buzz gets created, and those not involved ask why they're not invited and how they can participate.

How to Draw Up a Plan for a Successful LMR

Before attempting a limited market release, you have to create a plan and a protocol. During this process, you will be asking yourself and your team the following questions:

1. What's our objective, and what are we trying to prove and gather data on?

2. How are we going to execute this LMR?

3. What is the scope of the LMR?

4. Whom will we target and why?

5. How many units will we manufacture to satisfy the scope of the LMR?

6. What do we hope to learn from this exercise?

7. How will we measure results?

8. What is the beginning and what is the end point of this exercise?

Why the Complaint File Is an Innovator's Best Friend

This can be the most difficult time in the development process. You have been working hard to develop a beautiful new solution, and then you go out and hear bad news from your users. Criticism is always hard to hear, but that unwanted and painful criticism is actually more valuable than praise. Listening to the critics gives you an opportunity to learn and make changes that will only make your product better, easier to use, safer, and more effective over the long haul. Customer complaints, when taken seriously, will lead you to your next-generation product version and also tell your customers that you are listening to them, which is very important, especially today. This is the way you earn loyalty and respect from your customers. A partnership forms during an LMR from which you can create followers and helpers.

Not All Feedback Is Good Feedback

When you're all in, it's hard to tell the difference between constructive criticism and destructive criticism. Constructive criticism is offered in an attempt to help you make your product better. Destructive criticism serves a different agenda and has ulterior motives, like ego, for example. Remember my story in chapter 6 in the "Stop Talking and Listen" section. That was my ego getting the best of me.

It's critical at this stage to be able to take in all the feedback and then prioritize it in a way that is in line with your marketing and design specifications. That's your plan, so take all input in stride, and align it with customers' inputs. If the criticism is personal or without merit, you'll know, and it can be discarded. I do believe that those involved in the project in a direct or indirect role will have bought into the LMR program, because they have a vested interest in making the product successful.

The Remarkable Role of the World's Fair

When the world of marketing was young and innovators couldn't just hop off to Buenos Aires or Paris or Istanbul the way I do today, all releases were limited and generally took place at the World's Fair, where inventors and entrepreneurs from every nation would gather every four years to display their latest innovations, get feedback, interest investors, and build global buzz.

Emerging out of the French tradition of industrial expositions, the World's Fair as we know it began with the Great Exhibition of the Works and Industry of All Nations in London in 1851. It was the inspiration of Queen Victoria's husband, Prince Albert. It was intended to be the first international exhibition of manufactured products. It was a huge success, and from that point forward, new products that appeared at the World's Fairs were able to influence society, art, education, international trade, medicine, and manufacturing much more quickly. Suddenly, there was a world stage where innovations could be demonstrated and experienced in front of an audience of their peers as well as the general public. It was the first opportunity for inventors and brands to LMR their concepts, get valuable feedback from consumers, and test if their ideas were going to fly.

Almost every great invention of the last two centuries got its start at a World's Fair. Alexander Graham Bell introduced his telephone at the Philadelphia World's Fair in 1876. The Eiffel Tower was built as an entrance to the World's Fair in Paris in 1889. George Washington Gale Ferris Jr. debuted his Ferris wheel at the Chicago World's Fair in 1893. Josephine Cochran also showcased her new invention there, the dishwasher. X-rays were first demonstrated at the St. Louis World's Fair in 1904, and so was the very first American ice cream cone. Color TV, IMAX, and even the Ford Mustang all had the opportunity to change the world because of the World's Fair, the world's first limited market release platform.

Evaluating Your Results and Success Criteria

In my business, I need a success rate of nearly 100 percent, because lives are at stake. If a product doesn't work, very often the patient can get injured, or worse, they die. So, our success criterion is 99 percent efficacy. I won't get into a dissertation on statistics in this book. It's an understatement to say we require reliability and confidence in the products that we develop. Reliability and confidence—what's that mean? How reliable are your products? How confident are you about your reliability? Reliability is a measure of how well a product will perform under a certain set of conditions for a specified amount of time. Many times, reliability will be stated along with a minimum confidence level.

Not all inventions bear this heavy burden, but it's important to determine what you expect from your device and then measure how closely you get to meeting that goal every single time. And what kind of success ratio are you comfortable with? Are you hoping for a 100 percent result, like I do? Or are you comfortable with 98 percent? What numbers define success, and what numbers define failure for you? These are called your risk assessment criteria. The risk assessment of a product's failure rate must be understood and determined prior to launch. Then confidence and reliability intervals are established that determine how many units need to be tested to meet that criteria. A confidence and reliability ratio of 95/99 or higher, 99/99, could be required. Check the regulations in your industry and meet them.

Your success criteria also have to match the marketing and design specifications that you set out at the beginning. In my business, for example, our success criteria have to match the validation we submitted to get our regulatory approval. You can't raise the bar now, or lower it.

The other criterion for success is how closely your LMR results match your initial VOC results when you began this process. Did you really meet the customers' needs? Did you deliver what they were asking for? Did you genuinely solve a problem as you intended? Does the product function as you had intended initially? Can you make the margins you forecasted? Was the product easier or harder to build than you expected? Did your LMR go according to plan? Were you able to produce enough units to meet demand and with acceptable gross margin?

All of these questions will go into your evaluation of whether or not your LMR was successful. If your LMR satisfies your predetermined success criteria, then you are ready to plan a sales and marketing campaign and execute a full commercial launch.

CHAPTER 8

THE XTREME STRESS MANAGEMENT GUIDE FOR ENTREPRENEURS AND INNOVATORS

B eing an entrepreneur is probably one of the most stressful occupations one can choose. When you're in unmapped territory, all the decisions are yours, and that is a lot of pressure. When you make the right decision, there's a lot of glory. When you don't … well, there's a lot of the opposite.

The stakes are even higher when you take on investment dollars. At that point, you assume fiduciary responsibility for the company. The moment you take someone else's money and tell them that you have a plan to take their investment and return a multiple greater than what they gave you, the stress will inevitably kick in. Anything can go wrong, and you need to expect that some things inevitably will do just that. Mistakes will occur. Shit happens. However, when you have assembled the right team, those mistakes or setbacks become opportunities to make your solution better than you originally anticipated. Believe me, it happens. I've seen it. Failure is part of success.

Pressure Points for Young Innovators

These are the areas that can create the most stress for an innovator or entrepreneur. Knowing about them in advance can help you recognize them when they come along and cope with them more easily. I am enumerating them here so that you know that when those waves of doubt hit you, you're not alone.

1. **Pitching**

   Pitching to a potential investor creates stress. "Do they understand my story? I believe in my idea, but do they? What happens if nobody sees things the way that I do? What if I'm wrong?" These are questions that all young innovators or entrepreneurs ask themselves, so don't worry about those doubts when they come along. Everybody has them. Just stand your ground, rely on your instincts and your research, and make friends with even negative feedback. Nobody gets it right the first time out of the gate. So learn from your detractors, and grow from your doubts.

2. **Development**

"Will the solution we are creating work?" Once the investment is made, you've got to make the solution work. You can't quit now. You need to figure it out. This isn't a grant or free money you won in the lottery. It's someone else's hard-earned dollars and you have promised them it will realize a return. When you run into road-blocks, remember what got you this far down the road and find a way around them. Someone else believed in the idea and solution enough to invest in you, so trust their instincts when you are doubting your own.

3. **Recruitment**

"Did I hire the right people? Is this person going to be right for the team's chemistry and culture?" There are lots of smart people, and the finalists for my team are always smart and smarter than I am, but that's not the only thing that I look for in a team member. I want to know that they can communicate effectively and make our culture better. "What am I going to do if I have to get rid of someone? How will I replace them?" These are all questions that all young entrepreneurs ask themselves. This is why I strongly encourage every business leader to develop team rules and then let those team rules govern their hiring and firing practices. This will greatly reduce your stress level, because you can rest assured that if you follow your core values, you will wind up with a winning team.

4. **Competition**

"Is someone else working on the same solution that I am? Will I manage to be first to market, or will someone beat me? Will a bigger company steal my idea? What if I fail my regulatory tests? What if my raw material supplier tells me there is a delay or a cost increase?" When you are worried about issues surrounding competition, you can rely on the research that you did ahead of time to reassure you that you are in a good position from a market standpoint. Advance research and knowing the competition goes a long way toward answering these nagging questions and alleviating your fears.

The Entrepreneurial Lifestyle

When you're all in, your work schedule is going to disrupt your personal life and your commitments to your family and your friends. There's a dramatic downside to not being home, and it can put tremendous stress on your kids and on your wife, husband, or significant other, who will need to take on burdens that would otherwise be shared.

If you let it, being an entrepreneur and building a company or companies can really get in the way of forming and maintaining these essential bonds in your personal life. Without clear communication and understanding, you won't make it through the innovation process and come out the other side in one piece. I know that from personal experience. I've been divorced twice.

I try to build my friends and my family into my daily routine as much as possible. I make it a point to call one good friend at least once a week. Since the COVID-19 pandemic, I've been involved in homeschooling my kids and Zooming happy hours each week with many colleagues from around the world. I may be more connected to friends and family now than before. I coach my kids' sports teams. I take long weekends with my kids and my girlfriend. I remain friendly with both my exes and coparent harmoniously. I'm on local foundation boards in Orange County.

Just because you are all in with your job doesn't mean you are excused from being all in with your family and community. If you're going to be a successful entrepreneur, you have to do both. You can't pick one over the other, at least not all of the time. I've learned lessons during my journey and will continue to do so in the future.

On the other hand, an entrepreneurial lifestyle can be enormously fulfilling and rewarding for you and your family and friends. You and they may get to travel to places you never imagined going or meet with extraordinary people you and your family and friends would never have had the chance to meet any other way. I've been nearly all over the world. I have friends in the Middle East, South America, Southeast Asia, Europe, Canada. And I mean friends, not just business colleagues. These relationships enrich my life, and my family's life.

It's important to find ways to manage stress so that it doesn't flame out into your personal life. Exercise, meditation, nutrition, laughter, friendships—maybe even taking a ballet class like Prince George—do go a long way toward alleviating tension, generating energy, and making you a more well-rounded and pleasant person. This isn't just going to happen by itself, though. Like with every other step in your life as an innovator and entrepreneur, you have to have a methodology in place to manage the fallout from your high-powered lifestyle.

I also make it a point to build stress-relief activities into the daily life of my team. We've gone bowling or to batting cages, golfed, hiked. We go to the movies as a team. We have an innovation suite in our office with video games. And we've done off-site experiential training sessions, which sometimes make people more stressed out initially, but in the end, they build in systems to help everybody manage organizational tension.

DAVE'S FAVORITE ANTISTRESS ACTIVITIES

1. Play with my kids.

2. Socialize with friends often—in groups and one on one. Other people really help me to stay connected.

3. I'm a foodie, so I like to try new foods and talk

with chefs. Most chefs are scientists/chemists at heart.

4. Walk/run and exercise each week.

5. Get involved in local groups and foundations such as SAGE & AHA/ASA. Stay connected with my community.

6. Read books.

7. Travel to new and old places.

8. Get together annually with a group of high school buddies to see the New England Patriots play in a new city.

9. I am a father of three. My father coached my sports growing up. I loved that. So I do the same with my kids, as an assistant coach. I usually take a little more control, though, when I am able to be there.

10. I'm a tinkerer and I like to experiment with science. Building volcanoes, launching rockets with baking soda and vinegar, flying drones. My kids share the same interest. We have a lot of fun.

Stress Is Just Energy in Reverse

I'm a scientist, so I spend a lot of time looking at things through the lens of the physical laws of the universe, and I find that they are applicable to a lot more than just the way solids and liquids interact.

They are also applicable to people and to businesses. Take the first law of thermodynamics for example, also called the law of conservation of energy. The first law of thermodynamics states that energy can neither be created nor destroyed; energy can only be transferred or changed from one form to another.

For example, turning on a light doesn't actually produce energy but rather converts electrical energy into illumination. This law further says that there are only two kinds of processes, heat and work, that can lead to a change in the internal energy of a system. Since both heat and work can be measured and quantified, this is the same as saying that any change in the energy of a system must result in a corresponding change in the energy of the surroundings outside the system.

So why am I talking to you about the laws of thermodynamics in a business book? Because innovation is all about creating energy. Innovation companies, therefore, have to be all about making sure that the energy they are producing gets used productively rather than negatively, because energy is never destroyed. It just gets converted into other things through heat and work. So it's there whether you like it or not.

Think about it like this: When you're a kid playing baseball and you swing at the ball and the ball connects with the bat, that creates a level of stress. That stress recoils, which is why the ball flies off the bat. The greater the stress, the farther that ball is going to fly. That's creating positive energy in the environment. A home run is a desirable conversion of energy into a win.

If I'm in a meeting and I feel stressed, I feel the energy in me building up and creating stress in much the same way that the bat and ball feel. If somebody says something that I disagree with or don't understand, that stress recoils and produces an impact in the room. If something goes wrong on a test, sales are down, or management is allowing injustice to occur in a workplace without addressing it, stress gets generated, and when that stress recoils, people do things like blow up at their colleagues or quit. Obviously, this is a stress response that creates a negative impact on the environment. But if you are able to understand the root cause of the problem, look at things a different way, and create something novel, that is innovation—that is a productive conversion of energy into results that matter. That is a light bulb turning on.

Energy is all around us, all the time. I've been hearing the word *energy* my whole life. Excitement is a form of energy, as is ambition or love. Stress is just another form of this very same energy, which is created whenever two opposing elements connect and bond. The question is, how are you going to convert that energy? Will it produce a win or a loss?

Use It or Lose It

When you're in a situation where energy is being created—for example, if you're in a meeting and you're not paying attention, and I'm sitting next to you and I am paying attention, I'm listening to the people around me—I can actually take that energy from you. I can feed off the energy you haven't used and redirect it to convert into a result that is in line with my needs.

As you are sitting in this meeting and not listening, I'm in the meeting aware of the energy that's available. And while you're checked out, scribbling doodles on your notepad, I'm going to take that energy and redirect it. I will walk away from that meeting with a good result, and you will walk away feeling stressed out, fatigued, or frustrated, because I've taken your energy. The point is that in order to use the energy in a room that's available to you, you have to be ready to receive it.

When you're a leader, there are lots of ways to remain closed off to the energy around you and turn all of that positive energy potential into negative energy and stress. Here are just some of them:

- Not being open to your team's ideas

- Telling people that you have a better way

- Telling people to just trust you rather than inviting their feedback

- Implying that you know better than anybody on your team

These are all ways that a leader can remain closed off to energy, lose that energy, and ultimately lose the team.

Rob's Dilemma

I have a good friend, Rob, who is a study examining what happens to a company's energy that gets blocked by a leader or leadership. Rob is in the same field as I am. He started a company about ten years back with a product that looked very, very promising. The company in which he was the CEO was acquired a couple of years ago by a bigger company, and they have been putting the pedal to the metal ever since to either take the company public or have a larger M&A. Everything was going hunky-dory for him, and he stood to be compensated even more on the deal, so I didn't understand why he was calling me so upset.

"Dave, I can't stand it anymore," Rob said without any opening pleasantries. This was very unlike him. Rob was nothing if not thoughtful and polite. "I'm telling you I never would have let this company be acquired if I knew this was going to happen. This process is maddening. I'm ready to follow the rest of my core team right out the door at this point."

"I know it can be dicey at this juncture, Rob, but this is how it works. You know that," I said reassuringly. "Everybody gets over-the-moon stressed during this process."

"Of course, I know that," he said, a little annoyed. "This is different. This is not just normal growing pains; this is incompetent leadership. The division president hired an ops guy who is killing us. Sucking all the air right out of the room. He thinks he knows everything; he doesn't want feedback or help from anybody. He's leaving so much talent on the table it should be criminal. And yesterday, while he's flying all over the world first class doing God knows what, he told my engineers we can't afford to send them to a conference in Miami. In *Miami*, for God's sake, Dave. I could put fifty of my guys up at the Marriott Courtyard in Miami for what he spends on one trip to Paris."

"Is he doing a good job scaling your manufacturing at least?" I asked, because that was important to understand. My general rule is that I don't care what people spend as long as they're within budget and producing results. As long as what they are doing turns into increased gross margin and reduced expenses, by all means, live it up, within reason. That's the golden rule at this stage of the game.

"We've got millions of dollars in back orders," Rob said. "That's twenty percent more than what it was when he came in to fix that problem." Rob was right. This was sounding really bad.

"Have you talked to your president?" I asked, because I wanted to find a way out for him. I didn't want him to leave still feeling so upset. But I knew what he was up against. If he left now, they'd crucify him, blame him for all kinds of things he didn't do, and he'd leave his own handpicked team high and dry. It would be a mess.

"I've tried to talk to the boss, and he knows what's going on, but he's staked his reputation on this guy in front of the board. And he's terrible at confrontation. So, this stuff just keeps happening, and I've got an A team of engineers dropping like flies over here. What am I going to do, Dave? You're the guy who knows this team stuff inside out. So tell me, am I totally screwed, or is there a way around this? I'm getting pretty sick and tired of walking around apologizing to people for stuff he's doing, and I'm not alone. Everybody is looking over their shoulder. It's absolutely toxic around here. I don't see any solution but to leave, but I know they will crush me if I do that."

I thought Rob had one more shot at turning things around, so we made a date, and I ran him through my foolproof method for turning all of that negative energy and stress into positive team spirit.

Dave's High-Octane Corporate Energy Cleanse

Teams mean trust. If there is no trust, there is no team. That's true in baseball and hockey, and it's true in companies. When you build a team, trust is the first keystone you put in place. You have to build a team and work with a team with that always in mind—inspiring and preserving the trust. When there is malfeasance, or injustice, or poor communication, when people are undermining one another, or ignoring each other's input, or not sharing the glory, trust gets eroded. This is when people start to quit on you.

Engendering trust requires that leaders bring every team member along through victory and defeat. That means telling your team when you're happy with their performance and telling them when you're not. You have to be willing to be honest, real, and yes, even vulnerable in front of your team so that they feel comfortable enough to be the same way with you. This is the only way to free up blocked energy that is turning into frustration and stress. And over the years, I have developed a method for cleansing the energy in a room and rebuilding that fundamental trust between team members and leadership.

I do my cleanse generally as an off-site meeting, and I devote a couple of days, allowing people to get comfortable with risk and vulnerable. It's facilitated by an independent consultant or two. Exercises are designed to get the attendees to talk about the things that get under their skin. This is a retreat about taking chances and being honest and sometimes exposing secrets about themselves, about expressing resentments and concerns, and maybe even arguing. I have found that being honest and vulnerable about the problems is a great way to apply pressure where it's needed and reach resolution.

> You have to be willing to be honest, real, and yes, even vulnerable in front of your team so that they feel comfortable enough to be the same way with you.

It's also important during this retreat to learn as much as we can about each other as people. People are put into scenarios where they talk about themselves, about their upbringing, about the things that have shaped who they are, whether they realize it or not. At every meeting I get a handful of people who are really willing to take a leap of faith and share personal accounts of their lives. There are always those who are unwilling to do this. Some people just aren't ready to share on that level. But I always come away with a new and solid bond with the people who were willing to share. They are on my team for life. The ones who aren't willing to share and be honest lose out and don't grow.

I know that this kind of exercise is not for everybody. But if you want to build or rebuild trust, the best way that I have found is to share things with the team that people don't know about you, intimate and personal things that have impacted the way you see the world or make decisions in your job.

The things that I have learned, that you will learn through this process about your teammates and that they learn about you, really impact the way you respond to each other on the job. So often, the things that have happened to us in our past impact the way we perform on a team. Understanding these early influences in each other, and in ourselves, can be a game changer when it comes to stress between coworkers.

For example, I had real difficulties communicating. I found my first mode of self-expression through sports. This is a key reason why I build companies the same way that coaches build baseball teams and why I stress communication first and foremost in my hires. This early experience is also why I feel it's so important to get to know the person behind the patter, and that has shaped a lot of how I run my teams as well. If you had judged me back in those days based on the surface, you would never have realized who I was or what I was capable of.

My success consultants, whom I've worked with for years now and often bring in to help me run my energy cleanse, tell me that you are the way you are today because of what happened from birth to eight years old. Your experiences, your vulnerabilities, the stressors in your life can all be traced back to those first eight years. Maybe your parents got divorced, maybe you were a foster child, maybe you were a favorite child, maybe you were a black sheep, maybe you were rich, maybe you were poor, or maybe you had a speech impediment like me.

Who knows unless you share these things? But when you are willing to share these experiences with your teammates, it makes it easier for them to understand why you are the way you are, and that builds trust and eases stress.

This cleanse is about getting to know each other as people. This isn't about how to be a better manager or how to get the most out of your staff. It's about learning how to work together and win. And the results can be truly remarkable. I don't know whether or not it worked for Rob. Only time will tell, but I've started and run several successful companies, and it sure has worked for me. This is also training that people on my team are more comfortable doing on their own versus in the company team environment. Getting vulnerable in front of coworkers can be too much for most people. It can be easier in front of strangers. It takes work. I don't care how you do it. Just do it!

Money Is Energy Too

Hoarding money is the same thing as hoarding energy. When you are withholding necessary funds from the team in order to spend in less crucial areas that are important to you personally, you are withholding energy, and that is going to show up in your team's productivity. Take

Rob's leadership team member, for example. He is flying first class while treating his engineers like mushrooms in a dark room. He is unwilling to spend resources to send his engineers to conferences that would benefit them and the whole team, but he is willing to spend money on luxury travel. And so, ultimately, he will lose his engineers and maybe even his colleagues because he is misdirecting energy by misdirecting money.

This is a classic example of what happens on a team when you do not spread the wealth. When you hoard money, you block energy, and blocked energy translates into stress. You also engender mistrust, which strikes at the core of your team. When there is injustice, when priorities are personal rather than collective and the team senses this. When you speak truth to power and nobody listens—this is when you lose teams and when companies fail.

Do the Right Thing

As Abraham Lincoln once famously said, "You can fool all of the people some of the time, and some of the people all of the time, but you can't fool all of the people all of the time." Just ask Bernie Madoff. He had all the right credits on his résumé, all the right alma maters, zip codes, clients, and friends all over the world. Many of them were royal; many were complicit. Corruption takes a team as well, and Bernie had a great one. If anyone should have been bulletproof, it was Bernie Madoff. Yet it took just one guy, Harry Markolpolos, five minutes with a sales graph on a marketing brochure to understand that the whole thing was a fraud. Everybody gets caught eventually, but until then, a lot of stress gets generated, and a toxic environment erupts.

When you're in a field like innovation, where stakes are high and the land is uncharted, you run into corruption. I've seen it many times. Injustice flourishes for a time, but companies with malfeasance never endure. Because injustice erodes trust, without trust there is no team, and if there is no team, there is no innovation.

Postgame Rules of Success from Coach Dave

Whether you invented the triple Frisbee and it was a huge hit or a new embolization coil like I did, or even if you tried to reinvent the wheel and your effort has failed, the skills you develop during an entrepreneurial process will put you in good standing throughout your life, no matter what you choose to do next.

Learning to think like an entrepreneur means you have realized you can shape your life any way that you wish. It's up to you. That's powerful knowledge. When all options are on the table, which road do you choose? Where do you go from here? A good answer is by taking a look at what brought you to the entrepreneurial table in the first place—an ability to innovate around problems and find new solutions.

My road began when I was playing college baseball and I got injured. I knew after that injury that I would never play great. My and my dad's dreams of a future in major league baseball for me were over. To make matters worse, my grades weren't great either, because I had been partying too much. Hey, I was a jock. If I played well, who cared what my grades were, really? Now everything had changed. I was disappointed and worried that my dad would be disappointed in me too.

But my dad was a seasoned innovator by that point, and he moved on to plan B seamlessly and without looking back. He told me to get a summer internship and then go and do a co-op program at ULowell, or at a university like Northeastern, and phase into medical plastics. My dad set me up with a summer internship right away at an injection molding company in a town nearby, and it really was an eye-opening experience. I'll tell you one thing, I started to take my grades a lot more seriously, because I never wanted to work at another injection molding company...EVER!

That internship made an impression on me in many ways and really helped set my compass. I remember that one of the founders, let's call him Steve, drove a brand-new Jeep Wagoneer, which was the hottest SUV you could get at the time, and the CEO had a brand-new Mercedes every year. But the guy who ran the plastic injection floor … he drove an old Plymouth.

I was starting to get the idea that founding my own company might be a good direction for me. I liked the idea of a new Mercedes every year, I have to admit. At that time, it was about lifestyle, but then my grandfather got sick, and my thoughts about becoming an entrepreneur became more purposeful. I wanted to invent something that could have saved my grandfather. I wanted to make a difference in the world.

I talked to my dad, and he told me I couldn't go wrong with medical devices, because people would always get sick and need care, and new technology would always be needed to save lives. He said that he knew a guy, Doug Waterman, who worked for a catheter company. I explained to my school that I wanted to do a co-op program there. They told me that they had a better idea. There was a new plastic car being developed at GM called the Saturn and that even though my GPA was too low for their star students program, I was the most hands on, mechanically astute guy they had. The director said he wanted to nominate me for the Saturn team.

I told him that I wanted to go into medical devices, but he explained that this was the first all-plastic car. It would change the industry, maybe even make plastic history. They needed guys like me who knew how to take plastic molding to the next level.

So, I took the interview, and I flew out to Troy, Michigan, in the middle of December. It was, of course, beyond freezing, even for a New Englander like me. I saw the operation, and oh my God—the molding machines were the size of a garage. They were huge, monstrous things, and I was thinking, "Wait, I'm a hands-on guy. I don't know how to work with machines on this scale, and I don't want to learn how either. And I definitely don't want to live in Michigan in December."

Just looking at those enormous machines in the middle of a Troy winter told me I was in the wrong place. This wasn't a lifestyle

I wanted, this wasn't the climate I wanted, and it wasn't the work I wanted. Even though my university had stuck its neck out and given me a star recommendation despite my dismal GPA, I had the courage, or maybe just the chuzpah, to follow my instincts and turn that job down. And the rest, as they say, is history.

That decision, while I didn't know it at the time, was the very first entrepreneurial decision that I made on my own behalf. It led to a career developing products in the neurovascular space that literally saved lives and improved the quality of life for many others. Plus, it's given me a lifestyle that I enjoy in a place like Orange County, California, where it's warm and sunny. I really like my life, traveling the world speaking to other people in my field when needed, being able to spend time with my family, and starting companies, doing M&A, rinse and repeat. I really am living my dream, and I'm very grateful for that.

I am not sure whatever happened to the Saturn. Oh yeah—that car is no longer being made!

I bring up this story because it is this instinct that you can do it yourself and find a way to have it all, to hack the old way of doing things, that drives all innovators everywhere. We all share that instinct. But there are a lot of rules to learn along the way in order to sustain your choices.

Today, I am in the business of helping innovators to found and establish successful businesses based on their great ideas. I help them fill in where they are not strong and get their products to market much more quickly and successfully than they could have accomplished on their own. So, I've learned a few things along the way about building generational champions. Here are a few of my postgame rules for long-term success. Use them as a guide as you move forward into new and untrammeled highways.

And if you have a great idea that's in my lane, call me.

Follow your intuition, but back it up with action

Actions speak louder than words. Everybody thinks they have a brilliant idea. Very few of them ever hit the market, because there is a lot of legwork involved, right from the very beginning. Today, I'm teaching doctors who have ideas for new devices how to file patents, how to form C Corps and LLCs—stuff they never learned in medical school. There's no way around it. If you want to talk the talk, you have to walk the walk.

Know what you are good at, and stick to that

If you're great at team building and managing people, excellent. If you're a talented networker, then great—put that on your to-do list. The more you can accomplish on your own, the better off you are. That's how I found my niche. I'm an engineer who knows how to have a conversation. But remember, nobody is great at everything. Take an honest inventory of your strengths, and cultivate those. Don't waste time on the stuff you aren't good at.

Know what you don't know

While you're taking an honest inventory of your strengths, also take an honest inventory of your weaknesses. Don't try to do things that you're not good at; it's a waste of time. Bring in people who are proficient, continue to focus on your strengths, and always maintain humility. Honor those who can do what you can't, and compensate them generously. They are invaluable to you now and into the future. You have to have key people in key roles, and when you find them, hold on to them.

Always be a student

Always be open to learning things. This is at the core of innovative thinking, and this starts with being humble. You can't learn from others when you think you know everything already, and this is a mistake a lot of entrepreneurs who labor in a bubble make. Stay in touch with your teammates and with your industry. Stay abreast of what's up and coming and what you don't know anything about. Make connections with people who do. Never assume you know it all. Because you don't.

Everything you learn is valuable

Understanding this is very important for an entrepreneur. Every bit of wisdom that you gain is important, even though you may not know how at the time. You may use the information now, a year or two later, or even ten years down the line. You never know when one small piece of information will become crucial. Information is power. Hang on to every thread of it.

Know how to tell when you're on the wrong track, and adjust

If you're not finishing a task, if you're not showing up on time or hating your job, get out. When you're not getting along with your colleagues and coworkers, when things start to go south with your family, you are on the wrong path, and you will not succeed. It's important to recognize these signs first and act on them. Things will only get worse.

Patience, please

There are no shortcuts. Innovation takes time, and so does compliance with the regulations that govern your field. Do it right; follow the law. Trying to race through anything will only lead to lost time down the line. So, don't be impatient; be thorough, and don't fudge on the law, because it just isn't worth it.

Don't be too afraid of lawsuits

Lawsuits will happen when you're breaking new ground, and tossing around amounts like $100 million inevitably makes you a target. As long as you haven't cut corners, have done your job thoroughly, and have obeyed the law, there is nothing to be afraid of. Even if people come after you and your name is trashed on Google, you will ultimately prevail. If you get deathly afraid of nuisance lawsuits, you won't get anywhere as an entrepreneur. You need a thick skin and the confidence to know you are in the right. This confidence comes from knowing that you did the right thing at every turn. When you're in the wrong, litigation is terrifying. When you're in the right, you want your day in court.

Be a good boss

Payback is a bitch. In every industry, it's ultimately a small world at the top, and what goes around comes around. Don't make enemies by accident. Be a fair and generous employer, and you will be rewarded with loyalty.

Be cost conscious

When somebody spends extravagantly and blows through budgets, there's always going to have to be more money. This is a losing track. Don't throw good money after bad.

When opportunity knocks, answer the door, and come as you are

My entrepreneurial career began with a phone call from a recruiter, who said, "Hey, a friend of yours gave me your name. He has another friend who is a doctor, and we need you as an engineer to develop a product in the neurovascular space." And I said, "Okay." I really didn't know what the hell they were talking about when they told me they wanted to develop an embolization coil for aneurysms, but I said, "I can do that." Two weeks later I quit my job, signed my new contract, and began my life as an entrepreneur. When opportunities arise, be open to them.

Know where and how you work most productively

When I began my career as an innovator, I had a very flexible lifestyle. I worked at home. I could take my kids to school. After working three or four hours, I'd go to the driving range and hit golf balls. I met friends for lunch. I'd come back, work another three hours, go pick up my kids at school, then have dinner with my family. They'd go to bed. I'd work again from nine to midnight. It was entirely agile. Complete freedom.

After about three months, I couldn't stand it. All that freedom felt like confinement. I needed a workplace, I needed to get out of the house and so we built an office right in our competitor's backyard on the other side of the country. When you have the power to design your own work environment and your own schedule, pay attention to where you are most productive, how you are most comfortable, and structure an environment that you can enjoy long term.

Travel

One of the wonderful things, for me, about being an innovator is that the community of innovators is global. People all around the world are trying to solve the same problems that I am, so I make it a point to go and hear about how different people and countries approach the problem. In the process, my network has become global. I have seen and made friends in some of the most beautiful and inspirational places in the world. I have been to Israel ten times. I've hiked the Great Wall of China. I've been to the Brandenburg Gate in Berlin, and I've been to a Super Bowl, the Tour de France, and the Taj Mahal. Travel has enriched my life more than I can say.

I never would have done any of these things if I had just stayed in my job as an engineer at a big company. Take advantage of the global community in the industry you innovate for and the great opportunities for travel it offers. This will make you a better visionary, a better innovator, a better human.

Build generational wealth

When MindFrame was acquired and I had my first big chunk of cash, I realized that I suddenly had assets that I needed to protect if I wanted to pass that wealth along to my family and keep it safe from the many marauders that one encounters as an entrepreneur. What most people in that situation would do, I wanted to do—just go out and buy the new car, buy the boat—and I did my share of that. But fortunately I got into the hands of a good financial planner who helped me realize that I had to manage my assets if I hoped to generate reliable stability for my family and build generational wealth.

This planner helped me set up a Master Asset Protection Plan (MAPP), which protects me and my family, and my money, and still allows me and my family to live the lifestyle of my choice. It has given me great peace of mind and allowed me to focus on my goals in a new way. I cannot recommend this practice more emphatically, because it can make the difference between chasing dollars and chasing dreams.

Everyone's MAPP will differ based on their individual goals, but the philosophy is essentially the same. A MAPP is a plan to manage your assets so that you can operate, maintain, and renew the assets in the most cost-effective manner possible while maintaining your level of service and allowing your assets to appreciate.

No matter what your level of wealth at the moment, as you move forward as an entrepreneur creating companies and building wealth, it's a good idea to find a financial partner and attorney who can help you set up a protective castle where you can safely store all of your valuables.

Here is a quick overview of how you can begin to think about a Master Asset Protection Plan for yourself and your businesses.

1. **Define your assets.** Identify all of your assets. This means make a list of all items, physical, digital, or intangible, that have value to you and your business. This can be anything from a house to digital content to a brilliant idea you intend to exploit. Try to explain the value these assets have now and into the future.

2. **Take inventory of your assets.** Count, organize, and place value on the assets you have defined.

3. **Figure out asset demand.** The point here is to determine how much value needs to be created by your assets. This combines shareholder value creation demand, customer product demand, and regulatory requirements. This information will help you determine a level of service, which here refers to how much, how frequently, or how

many times you will have to use each asset to create value.

4. **Determine the cost of each asset over its lifespan.** Determine what your assets will cost to maintain at their current level of value. For example, if your asset is a house, how much money will you need to maintain, service, or repair that house over its lifetime?

5. **Reduce risk.** Measuring asset risk gives you a way to identify and mitigate your exposure to risk and put systems in place to replace risky assets or put backup assets in place to cover the costs. What is the chance that your assets will break down, fail to provide service, or fail entirely? What is the cost of that failure? How likely is it that you could be sued? What is the cost of protecting your assets from that threat?

6. **Minimize cost.** Look for ways to optimize the efficiency of acquiring, managing, maintaining, moving, or using your assets. This will mostly depend on the nature of your business's assets. Figure in the cost of maintaining assets over a long period of time, and decide whether it would be more cost effective to just replace. Alternatively, you could find more efficient methods of moving your assets around as you need. Consult with your management team to brainstorm cost-reduction ideas.

Taking an inventory like the one above will help you to have a more substantial discussion with a financial planner. There are also many software programs now that will help you at the beginning to set up your assets in a way that will protect them and help them grow.

MY BLUE MARBLE

One of the best things about being an entrepreneur is that you get a lot of freedom to customize your life and are able to shake it up whenever you want. As an innovator, I can write my own story. And so can you.

Granted, reinvention is scary. Change and disruption, trying new things is always challenging at the top. It's like the moment right before the roller coaster stops going up, and you are waiting for that split second of terror mixed with exhilaration before you start the trip down. If you're anything like me though, when you get to the bottom, you're always ready to go again.

As I finish the writing of this book, I'm in the process of yet another major shake-up and reinvention period in my life. I sold my last company, I put in my time helping manage the transition, and I have now decided to embark on the next big thing. I've started a new company called Quantum with some friends of mine whom I really enjoy. We all have different talents and different backgrounds, but we think alike, we enjoy similar things, we all are committed and work hard, and we all share similar values and goals.

Quantum is a custom business model called a venture studio, which is an organization that creates startups, typically by providing the initial team, strategic direction, and capital for the startup to reach product-market fit. The venture studio model is different than a traditional startup accelerator. Accelerators provide what is typically a twelve-week program and initial seed funding ranging from $25,000 to $125,000 in capital in exchange for a set amount of equity. The program provides training and guidance to validate the business. At the end of the program, the accelerator has a "demo day" where investors can learn about the latest startup batch and hopefully invest in the accelerator's portfolio.

Accelerators tend to focus on spreading small amounts of capital across a wide array of startups with the expectation that most will fail. Our venture studio will take a slightly different approach, focusing more resources around opportunities that we identify as ripe for a startup to capture. Unlike accelerators, we don't typically accept applications for new portfolio companies, as the venture studio's strategic insight and ability to select opportunities is a part of the value it brings to its investors.

Venture studios came into existence in the late 2000s after a handful of seasoned tech entrepreneurs with successful exits found that they had a lot more than one startup idea to pursue. Building a company from scratch is hard, usually requiring a minimum ten-year commitment. This wasn't working for serial entrepreneurs like myself who really enjoy the early stage of the process—discovering and proving out a new business model and building out a new organization—more than the process of growing a proven business with established processes. These are two different skill sets, two different passions, and if I've taught you anything in this book, I want to be sure that you understand that it's good to know what you like, and what you're good at, and then just do that.

I chose the venture studio route because I have like-minded complementary (business development, engineering, manufacturing, regulatory, and quality) business colleagues who are focused on a specific technology sector called interventional radiology. Together, we can manage and develop several products in parallel versus doing just one innovation at a time. This business strategy is much more efficient and effective for investors and our customers in the market who want to be part of bringing diagnostic and therapeutic devices to the market and improving their patients' quality of life and saving lives.

I like this kind of approach to innovation, which in many ways mirrors the way the inventors and artists throughout history have modeled their incubators, stretching all the way back to Renaissance Italy. And I'm Italian, so I like that idea. I want to create a grand innovation house where we do things in our own signature style. I want to be a studio that stands for mastery, creativity, collaboration, specification, and mutual support. I want to build a winning team that will last beyond me, and I want to do it with the people of my community who aren't afraid to shake it up with me, to put it out there on the wire, and to see if we can find a better way.

> I like this kind of approach to innovation, which in many ways mirrors the way the great inventors and artists throughout history have modeled their incubators, stretching all the way back to Renaissance Italy.

When I was younger, I read about something they call "Blue Zones." Blue Zones are the places in the world where people live the longest and the healthiest lives. This really impacted me, and I have often thought over the years that I would love to live in a Blue Zone, but those cultures and places just seemed so far removed from my own life in Orange County, California. These Blue Zones all have some traits in common, no matter what culture they emerge from, no matter where they are located in the world. Here's what the Blue Zones all have in common:

1. **Life Purpose:** People in Blue Zones wake up every day for a reason and with a sense of purpose.

2. **Eat Earlier in the Day:** People in Blue Zones eat their smallest meal in the late afternoon and don't eat after that.

3. **Strong Community:** Almost all of the people in Blue Zones who have reached the age of one hundred report being part of some faith-based community.

4. **Good Relationships:** Blue Zone people put family first. They are more likely to keep aging relatives nearby or in the house. They also commit to a spouse or life partner and commit to spending quality time with their children.

5. **Natural Movement:** People in Blue Zones incorporate movement into their everyday life routines, either through growing their own food or manual labor. They also clean their own houses and cook their own meals.

6. **Relaxation:** People in Blue Zones don't have chronic stress, and relaxation periods are built into their normal routines.

7. **Eat Plant Forward:** People in Blue Zones eat a plant-forward diet and enjoy meat usually only once a week in small portions.

8. **Wine:** People in Blue Zones have a couple of glasses of wine every day.

9. **Find Their "Tribe":** People in Blue Zones live in communities with people who share similar values and lifestyles.

While I can't see myself returning to a lifestyle in a small village where multiple generations live together in like-minded and stress-free harmony—I am engaged in an effort to innovate a way to create my own version of a Blue Zone, which I'm calling my Blue Marble, right here in Orange County.

I want my Blue Marble to be purposeful. I want to benefit my industry by bringing more new products to market faster. I want to establish a neuro-vascular Silicon Valley right here in Orange County. I want to encourage and shape young talent in my field.

In my Blue Marble I want to work in a way that puts family and friends first. I want to work with people who feel the same way about that as I do. I want to work with people who are team players, who want to win games but also live life to the fullest. I want to spend my time with people who are willing to push the envelope, try new things, and go places that scare them. I want people in my marble who will play ice hockey with me, who have kids who will play soccer with my kids. I want to be more than a business culture; I want to be a community-centered business.

I want my Blue Marble to be a place where my kids want to grow up and stay, not grow up and leave as fast as their feet will carry them. I want my son, who just got his business degree, to come work with me and raise his family here in Orange County. If my daughter wants to go into horses or my youngest son wants to be a YouTuber, I want a community-centered business that is agile enough to accommodate that too. I want the diverse talents of the next generations to have the space and freedom to bring new roads to culture, innovation, and mastery that expand and enhance the studio in ways I can't even imagine or predict now.

I want my blue marble to be a place where great ideas, big ideas, are encouraged, supported, and executed with efficiency, imagination, and passion. I want to do this, because I can.

Now that you have had your first successful product launch, what can you do next?

What will your blue marble look like?

Find out more about my blue marble at:

https://www.treadstoneoc.com

.